Felix Rieper

On the behaviour of numerical schemes in the low Mach number regime

Felix Rieper

On the behaviour of numerical schemes in the low Mach number regime

An Analysis of the Dissipation Mechanism of Upwind Flux Functions on Different Cell Geometries

Südwestdeutscher Verlag für Hochschulschriften

Impressum/Imprint (nur für Deutschland/ only for Germany)
Bibliografische Information der Deutschen Nationalbibliothek: Die Deutsche Nationalbibliothek verzeichnet diese Publikation in der Deutschen Nationalbibliografie; detaillierte bibliografische Daten sind im Internet über http://dnb.d-nb.de abrufbar.

Alle in diesem Buch genannten Marken und Produktnamen unterliegen warenzeichen-, marken- oder patentrechtlichem Schutz bzw. sind Warenzeichen oder eingetragene Warenzeichen der jeweiligen Inhaber. Die Wiedergabe von Marken, Produktnamen, Gebrauchsnamen, Handelsnamen, Warenbezeichnungen u.s.w. in diesem Werk berechtigt auch ohne besondere Kennzeichnung nicht zu der Annahme, dass solche Namen im Sinne der Warenzeichen- und Markenschutzgesetzgebung als frei zu betrachten wären und daher von jedermann benutzt werden dürften.

Verlag: Südwestdeutscher Verlag für Hochschulschriften Aktiengesellschaft & Co. KG
Dudweiler Landstr. 99, 66123 Saarbrücken, Deutschland
Telefon +49 681 37 20 271-1, Telefax +49 681 37 20 271-0
Email: info@svh-verlag.de
Zugl.: Cottbus, BTU, Diss., 2008

Herstellung in Deutschland:
Schaltungsdienst Lange o.H.G., Berlin
Books on Demand GmbH, Norderstedt
Reha GmbH, Saarbrücken
Amazon Distribution GmbH, Leipzig
ISBN: 978-3-8381-0196-5

Imprint (only for USA, GB)
Bibliographic information published by the Deutsche Nationalbibliothek: The Deutsche Nationalbibliothek lists this publication in the Deutsche Nationalbibliografie; detailed bibliographic data are available in the Internet at http://dnb.d-nb.de.

Any brand names and product names mentioned in this book are subject to trademark, brand or patent protection and are trademarks or registered trademarks of their respective holders. The use of brand names, product names, common names, trade names, product descriptions etc. even without a particular marking in this works is in no way to be construed to mean that such names may be regarded as unrestricted in respect of trademark and brand protection legislation and could thus be used by anyone.

Publisher: Südwestdeutscher Verlag für Hochschulschriften Aktiengesellschaft & Co. KG
Dudweiler Landstr. 99, 66123 Saarbrücken, Germany
Phone +49 681 37 20 271-1, Fax +49 681 37 20 271-0
Email: info@svh-verlag.de

Printed in the U.S.A.
Printed in the U.K. by (see last page)
ISBN: 978-3-8381-0196-5

Copyright © 2010 by the author and Südwestdeutscher Verlag für Hochschulschriften Aktiengesellschaft & Co. KG and licensors
All rights reserved. Saarbrücken 2010

to my father

and to

Florian and Jasmin

Acknowledgements

I want to thank my thesis advisor Prof. Dr. Georg Bader for his continuing interest and support of my work. After a few years outside university he gave me the opportunity to come back to science and research at his Chair of Numerical Analysis and Scientific Computing. His trust in my capabilities and the great amount of freedom offered enabled me to work intensively on this thesis.
Special thanks to Dr. Sabine Roller for supporting my usage of the HYDSOL code and for raising my interest in the field of low Mach number flow.
I am grateful to Dr. Friedemann Kemm who always had an ear for my small problems and spontaneous ideas.
Prof. Dr. Rupert Klein gave me the opportunity to present my work in his team, which lead to numerous improvements in the thesis. Thanks also to Prof. Dr. Rupert Klein and Stefan Vater for the invitiation to the Doktorandenseminar and the fruitful discussions afterwards.
Thank you Prof. Dr. Ekkehard Köhler for the support in the field of graph theory.
For reviewing the manuscript at various stages I thank my father Karl-Otto Eschrich, Dr. Maren Hantke and Piet Schwarzenberger.

Cottbus, December 4, 2008

Abstract

This work is concerned with the study of two-dimensional inviscid flow. The main concern is the so-called *accuracy problem* of first-order upwind schemes in the low Mach number regime. The thesis is divided into two parts.

In a preliminary chapter the governing thermodynamic and mechanical equations are introduced. The Navier-Stokes equations for viscous flow are presented because the concepts of fluid viscosity and Reynolds number are later needed in the context of artificial viscosity of numerical schemes. The Euler equations are introduced along with a single-scale asymptotic analysis for small Mach numbers.

In the first part the behaviour of various first-order upwind schemes in the low Mach number regime is analysed with respect to a *one-dimensional* model case, where the two-dimensionality only remains in the shear waves of the local Riemann problem. The concepts of *numerical viscosity* and *numerical Reynolds number* are first introduced for the upwind method for the scaler linear advection equation. These concepts are then applied to Roe's scheme and the scheme by Harten-Lax-van Leer (HLL). The aim is to show that the *accuracy problem* can only be avoided, if all characteristic waves are resolved by the upwind scheme. Otherwise, the artificial viscosity on these waves is of the wrong order of magnitude $\mathcal{O}_S(\Delta x/M)$ and grows with decreasing Mach numbers, instead of being $\mathcal{O}_S(\Delta x)$. In the following chapter, various flux vector splitting methods are analysed in this respect.

The analysis presented is verified with a number of numerical results. Comparisons with analytical solutions give a direct measure of the accuracy of the scheme; to this end we use the flow around a cylinder. As standard test case the flow around a NACA0012 aerofoil is used.

The second part of the thesis is dedicated to *two-dimensional* flow. The numerical scheme of choice is based on Roe's approximate Riemann solver, which does not show the accuracy problem in the one-dimensional setting.

Part II begins with broad numerical studies, which path the way to the analysis by giving answers to the following questions. How does the numerical behaviour of the scheme change, when the cells in a structured, body-fitted grid around a cylinder are continuously transformed from (almost) squares, via trapezoids to triangles? How important is the grid structure to the *accuracy problem*?

Later we restrict the numerical investigations to structured grids with two different cell geometries: squares and rectangular triangles, which originate from the squares by adding a diagonal edge. On these types of finite volume cells the steady flow around a rectangle and the dissipative behaviour of an oblique contact layer is investigated. The moving Gresho vortex is simulated as an example of an unsteady flow simulation. All numerical studies suggest two things: firstly, the *accuracy problem* is linked to the momentum transport in shear flow and, secondly, this problem is absent if the finite volume cells are of triangular shape – irrespective of the grid being structured or unstructured.

The analysis begins with heuristic considerations. We investigate a special case of a divergence free flow with constant pressure and density on a Cartesian and a related triangular grid. Already here, the different form of the artificial viscosity term on square and triangular finite volume cells becomes evident. On *Cartesian grids* there is a decoupling of the artificial viscosity into horizontal and vertical direction. For nontrivial velocity fields the viscosity terms of order $\mathcal{O}_S(\Delta x/M)$ cannot vanish and cause the numerical error to grow for decreasing Mach numbers. On *triangular grids* there is no such decoupling: the artificial viscosity of order $\mathcal{O}_S(\Delta x/M)$ contains all jumps of the normal component of the velocity at the three cell interfaces. This suggests that the *accuracy problem* is absent because these jumps can completely vanish without constraining the velocity field to a trivial flow.

The thesis closes with a proof of this proposal for the Roe scheme for steady low Mach number flow. Using asymptotic analysis of the semi-discrete equations, it is explicitly derived that the *accuracy problem* is avoided, if the Roe scheme is used on triangular finite volume cells. The accuracy of the scheme is accompanied by a constraint: the normal velocity does not jump at cell interfaces. This constraint leaves enough degrees of freedom for the velocity field as is shown with graph theoretic arguments.

Zusammenfassung

Kern dieser Arbeit sind zweidimensionale reibungsfreie Strömungen. Das Hauptaugenmerk liegt dabei auf dem sogenannten *accurcay problem* von Upwind-Verfahren erster Ordnung im Bereich kleiner Machzahlen. Die Dissertation ist in zwei Teile untergliedert.

In einem einleitenden Kapitel werden die für diese Studie relevanten thermodynamischen und mechanischen Gleichungen eingeführt. Die Konzepte der inneren Reibung in Fluiden und der Reynoldszahl werden später im Kontext von künstlicher Viskosität numerischer Verfahren benötigt. Daher werden die Navier-Stokes Gleichungen für reibungsbehaftete Strömungen zuerst eingeführt. Im Anschluss werden die Eulergleichen und die Ergebnisse der asymptotischen Analyse für kleine Machzahlen vorgestellt.

Im ersten Teil der Arbeit werden verschiedene Upwind-Verfahren erster Ordnung hinsichtlich ihres Verhaltens in einer *eindimensionalen* Modellsituation untersucht, wobei die Zweidimensionalität nur durch die Existenz von Scherwellen in den lokalen Riemannproblemen bestehen bleibt. Das Konzept der numerischen oder künstlichen Viskosität und der numerischen Reynoldszahl wird am Beispiel des Upwind-Verfahrens für die skalare lineare Advektionsgleichung entwickelt.

Diese Konzepte werden dann auf das Roe-Verfahren und das Verfahren von Harten, Lax und van Leer (HLL) übertragen. Ziel dieser Analyse ist es zu zeigen, dass im Bereich kleiner Machzahlen ein Upwind-Verfahren alle charakteristischen Wellen auflösen muss, wenn das *accuracy problem* vermieden werden soll. Dieser Aspekt wird im Anschluss für diverse Fluss-Vektor-Splitting Verfahren untersucht.

Die Analyse wird mit zahlreichen numerischen Ergebnissen untermauert. Die stationäre Zylinderumströmung wurde als Testfall gewählt, da die analytische Referenzlösung der inkompressiblen Potentialströmung bekannt ist. Als weiteren Testfall untersuchen wir die stationäre Strömung um ein NACA0012 Flügelprofil.

Der zweite Teil der Arbeit ist *zweidimensionalen* Strömungen gewidmet. Als Flusslöser beschränken wir uns hier auf den approximativen Riemannlöser von Roe, der das *accuracy problem* in der eindimensionalen Modellsituation nicht aufweist.

Dieser Teil beginnt mit einer Reihe von numerischen Studien, die Hinweise darauf geben sollen, in welcher Richtung die Analyse anzusetzen ist. Für die Zylinderumströmung wird das strukturierte, körperangepasste Gitter derart transformiert, dass das Verhalten des Verfahrens beim stetigen Übergang von quasi quadratischen Zellen über trapezförmige bis hin zu dreieckigen Zellen ersichtlich wird. Vergleichende

Rechnungen auf strukturierten und unstrukturierten Gittern um den Zylinder sollen Aufschluss geben über den Einfluss der Gitterstruktur auf das *accuracy problem*.

Im Hinblick auf eine vereinfachte Analyse werden die numerischen Untersuchungen im Anschluss auf strukturierte Gitter mit zwei Zellgeometrien beschränkt: Quadrate und rechtwinklige Dreiecke, die durch Einfügen einer Diagonalen aus den Quadraten hervorgehen. Auf diesen zwei Typen von Finite-Volumen-Zellen wird die stationäre Umströmung eines Rechtecks sowie das dissipative Verhalten auf schräglaufende Kontaktunstetigkeiten untersucht. Der Transport eines Greshowirbels dient als Beispiel einer instationären Strömung.

Alle numerischen Ergebnisse legen zwei Dinge nahe: Erstens, dass das *accuracy problem* verbunden ist mit dem Impulstransport in Scherschichten, und zweitens, dass dieses Problem auf dreiecksförmingen Finite-Volumen-Zellen vermieden wird – unabhängig davon, ob das Gitter strukturiert oder unstrukturiert ist.

Die Analyse des *accuracy problems* beginnen wir mit einigen heuristischen Überlegungen, die von vereinfachten Bedingungen ausgehen. Dazu untersuchen wir den Spezialfall einer divergenzfreien Strömung konstanter Dichte und konstanten Drucks auf zwei einfachen Gittertypen mit quadratischen Zellen und mit den abgeleiteten dreieckigen Zellen. Dabei wird das grundsätzlich verschiedene Verhalten des Roelösers auf diesen beiden Zelltypen bereits an der Form der numerischen Viskosität deutlich.

Bei *kartesischen Gittern* beobachtet man eine Entkopplung der künstlichen Viskosität in horizontale und vertikale Richtung. Für nicht-triviale Strömungen können daher die Fehlerterme der Ordnung $\mathcal{O}_S(\Delta x/M)$ nicht verschwinden – der Diskretisierungsfehler wächst mit abnehmender Machzahl. Auf *Dreiecks-Gittern* gibt es keine solche Entkopplung. Die künstliche Viskosität der Ordnung $\mathcal{O}_S(\Delta x/M)$ enthält die Sprünge der Normalenkomponente der Geschwindigkeit von allen drei Zellkanten. Dies legt die Vermutung nahe, dass auf Dreiecksgittern das *accuracy problem* vermieden wird, indem diese Sprünge verschwinden – ohne das Geschwindigkeitsfeld zu sehr einzuschränken.

Die Studie wird mit einem Beweis dieser Vermutung für den Roelöser im Bereich kleiner Machzahlen im stationären Fall abgeschlossen. Mit Hilfe einer asymptotischen Analyse des semi-diskreten Roe-Verfahrens wird gezeigt, dass das *accuracy problem* auf dreiecksförmigen Finite-Volumen-Zellen vermieden wird. Man erhält jedoch eine zusätzliche Zwangsbedingung an das Geschwindigkeitsfeld: die Normalkomponente der Geschwindigkeit an einer Zellkante springt nicht. Mit Hilfe von graphentheoretischen Argumenten kann man zeigen, dass trotz dieser Zwangsbedingung noch genügend Freiheitsgrade für das Geschwindigkeitsfeld übrig bleiben.

Contents

1. Introduction   1
    1.1. Mach number and compressibility 1
    1.2. Reynolds number and dissipation 2
    1.3. Low Mach numbers in CFD 3
    1.4. Organisation of the thesis 4

2. Governing equations   7
    2.1. Equations of state for a perfect gas 7
       2.1.1. Thermal equation of state 7
       2.1.2. Caloric equation of state 8
       2.1.3. The entropy 9
    2.2. The Navier-Stokes equations 10
       2.2.1. Introduction to asymptotic analysis 12
       2.2.2. Asymptotic analysis for small Reynolds numbers 14
       2.2.3. Stokes flow around a cylinder 15
    2.3. The Euler equations 17
       2.3.1. The Strouhal number 18
       2.3.2. Asymptotic analysis for small Mach numbers 20
       2.3.3. Incompressible potential flow around a cylinder 26

I. Analysis for 1D: flux solver   29

3. Concept of numerical Reynolds number   31
    3.1. Introduction 31
    3.2. Physical and artificial viscosity 32
       3.2.1. The modified equation approach 32
       3.2.2. Reynolds numbers for numerical schemes 34
    3.3. Upwind methods for linear scalar equations 34
       3.3.1. Explicit upwind method (CIR) 34
       3.3.2. Implicit upwind method 35
       3.3.3. Implicit upwind method with large time steps 36
    3.4. Summary 37

4. The Roe scheme — 39
- 4.1. Characteristic equations — 39
- 4.2. The first-order Roe scheme — 40
  - 4.2.1. Spatial discretisation — 40
  - 4.2.2. Explicit time discretisation — 43
  - 4.2.3. Numerical results — 44
  - 4.2.4. Implicit time discretisation — 53
- 4.3. Summary — 54

5. The HLL scheme — 57
- 5.1. The two-wave HLL scheme — 57
  - 5.1.1. Spatial discretisation — 57
  - 5.1.2. Explicit time discretisation — 60
  - 5.1.3. Numerical results — 61
  - 5.1.4. Roe and HLL – a comparison — 68
  - 5.1.5. Implicit time discretisation — 69
- 5.2. HLLC — 71
  - 5.2.1. Transport of a shear wave — 72
  - 5.2.2. Numerical results — 73
- 5.3. Summary — 74

6. Flux vector splitting methods — 75
- 6.1. The van Leer splitting — 75
- 6.2. The Steger-Warming splitting — 76
- 6.3. The Liou-Steffen scheme (AUSM) — 78
- 6.4. Numerical results — 79
- 6.5. Summary — 84

7. Pressure decomposition — 85
- 7.1. Viscosity-Induced Pressure Fields — 85
  - 7.1.1. Asymptotically consistent schemes — 86
  - 7.1.2. Asymptotically inconsistent schemes — 87
- 7.2. Numerical results — 87

II. Analysis for 2D: cell geometry — 91

8. Numerical experiments — 93
- 8.1. Flow around a cylinder — 93
  - 8.1.1. Grid transformation: from rectangular to triangular cells — 94
  - 8.1.2. Structured vs. unstructured triangular grids — 95
  - 8.1.3. A layer of perturbation cells — 98
  - 8.1.4. Convergence study — 98
  - 8.1.5. Efficiency study — 100

8.2.	Flow around a square	102
8.3.	Low Mach number region	105
8.4.	Gresho vortex	105
8.5.	Inflow of a contact layer	109
	8.5.1. Steady entropy layer	109
	8.5.2. Steady shear layer	111

9. Quadrilateral grid cells — 115
9.1. Discrete divergence constraint 116
9.2. Cartesian grid 117
    9.2.1. Transport of momentum 118
9.3. Generalisation 120

10. Triangular grid cells — 125
10.1. Triangulation derived from a Cartesian grid 125
    10.1.1. Upper triangle 126
    10.1.2. Lower triangle 130
    10.1.3. Discrete divergence constraint 131
10.2. Dual grids of triangulations 133

11. Proof for special triangulations — 137
11.1. Constancy of the leading-order pressure $\mathbf{p}^{(0)}$ 139
11.2. Constancy of the first-order pressure $\mathbf{p}^{(1)}$ 140
    11.2.1. Internal cells 141
    11.2.2. Boundary conditions 147
11.3. Degrees of freedom for $\mathbf{u}^{(0)}$ 152
    11.3.1. Introduction 152
    11.3.2. Motivating examples 152
    11.3.3. Graph theoretic analysis 157

Summary and Outlook — 163

Appendix — 167
A.1. Roe's approximate Riemann solver 167
A.2. Difference operators in 2D 171
A.3. Rules for asymptotic expressions 173
A.4. Asymptotic equations for a perfect gas 174
A.5. Asymptotic equations of the Roe scheme 175
A.6. Software 184

Notation — 189

List of Figures — 191

Bibliography — 197

1. Introduction

Computational fluid dynamics (CFD) was for a long time rigidly divided in simulating compressible and incompressible flows. But a variety of important flow phenomena like atmospheric flows in meteorology, deflagration-detonation transition (DDT) in reactive flows or low velocity flows with large density variations as in tunnel fires are quasi-incompressible with weakly compressible effects. The simulation of these kinds of flow poses a number of problems and is a matter of ongoing research. Before addressing these problems, we elucidate some of the major concepts this thesis is build on.

1.1. Mach number and compressibility

To avoid misunderstandings, there are two distinct definitions of compressibility: in thermodynamics *compressibility κ of a material*, e.g. a gas or a liquid, describes the relative change of density ρ as a response to a change in pressure p:

$$\kappa = \frac{1}{\rho} \frac{\partial \rho}{\partial p}.$$

Note that there is no specification as to what the origin of the pressure change is. The *compressibility of a flow* measures the relative change of density due to pressure gradients originating in the flow (for a more general definition see [4] for example). Interestingly, this flow property is inseparably linked to the dimensionless number

$$\boxed{\mathrm{M} := \frac{u}{a}},$$

called *Mach number* – the ratio of flow velocity u and sound speed a. For perfect gases a is given by

$$a^2 = \frac{\gamma p}{\rho} \tag{1.1}$$

with the adiabatic index γ. We demonstrate the role of the Mach number with a simple order analysis. The magnitude of the pressure fluctuations in inviscid flow originates from two physical principles: energy conservation and inertia.

On the one hand, the interaction of the flow with an obstacle leads to a local acceleration of the fluid. Bernoulli's principle, cf. [61] for example, demands that the resulting pressure fluctuations are of the same order of magnitude as the kinetic energy which causes these perturbations. The magnitude of the relative change of density for a flow around an obstacle is given by the proportionality relations

$$\frac{\Delta\rho}{\rho} \sim \frac{\Delta p}{p} \sim \frac{\frac{1}{2}\rho u^2}{p} \sim \frac{u^2}{a^2} = \mathrm{M}^2 \, ,$$

where we have used from left to right: a standard compressibility law for a fluid, Bernoulli's principle, the speed of sound in gases $a^2 \sim p/\rho$ and the definition of the Mach number.

On the other hand, inertia forces f in vortices are proportional to the square of the flow velocity: $f \sim \rho u^2/l$, where l is the size of the vortex, resulting in a similar relation for the relative change of density in vortices:

$$\frac{\Delta\rho}{\rho} \sim \frac{\Delta p}{p} \sim \frac{\nabla p \, l}{p} \sim \frac{f \, l}{p} \sim \frac{u^2}{a^2} = \mathrm{M}^2 \, .$$

In other words, the *square* of the Mach number is a direct measure of the compression caused by the flow and allows a classification of flow regimes: if the square of the Mach number is smaller 0.1 we call a flow *weakly compressible*; if it is larger we have *compressible flow*. In some applications small compression effects are of minor interest – in this case the flow is assumed *incompressible*.

Historically, in the latter situation the density was explicitly set constant, which lead to the *incompressible flow equations*. These equations were later shown to be the limit equations of the compressible flow equations by Klainerman and Majda [26] with a single-scale asymptotic analysis. Klein [27] introduced an acoustic length scale to investigate the influence of long wave acoustics on the flow with a multi-scale asymptotic analysis. Long wave acoustic occurs in flows with energy sources as in combustion or meteorology. Note however, there are *incompressible flows* with variable density treated, for example, in [47, 1].

1.2. Reynolds number and dissipation

The Reynolds number Re is a dimensionless number, which enters the equations of *viscous* flow, when all physical quantities are made dimensionless by a scaling. With the scaling of the momentum equation the ratio of convective transport to dissipation:

$$\boxed{\text{Reynolds number} = \frac{\text{magnitude of momentum convection}}{\text{magnitude of momentum dissipation}}}$$

is introduced into the equation. In most engineering contexts the Reynolds number is used to identify the transition from laminar to turbulent flow.

Two limiting cases are of interest in this thesis. For high Reynolds numbers the momentum dissipation can be neglected; we then have *ideal* or *inviscid flow*, where the pressure is governed by Bernoulli's law with $\mathcal{O}_{\mathrm{S}}(\mathrm{M}^2)$ fluctuations. In the case of low Reynolds numbers, the momentum transport is dominated by dissipation. As a consequence, the pressure field is no longer determined by Bernoulli's law, but by the gradient of viscous forces causing pressure variations of order $\mathcal{O}_S(M)$. Flows of this kind are known as *creeping* or *Stokes flow*.

1.3. Low Mach numbers in CFD

The computation of low Mach number flows faces a variety of problems. Schemes for *incompressible flows* such as Chorin's projection method [10] exclude any density variations right in the model equations and are inadequate for simulating flows with relevant density variations, such as thermally driven flows. Yet, a variety of new schemes [23, 40, 41, 11] can be interpreted as extensions of the classical schemes for incompressible flows. They adopt the pressure correction approach but allow, at the same time, for density variations. Extensions of incompressible solvers using multiple pressure variables (MPV) based on an asymptotic analysis have been developed in [46, 44, 39].

Schemes designed to calculate *compressible flows* – originally with focus on capturing shock waves in high speed flows – encounter three major problems in low Mach number flows:

1. The wave speeds of acoustic and flow phenomena are of different orders of magnitude – their ratio is measured by the Mach number. Low Mach numbers slow down the calculation of phenomena on the time scale of the flow such as heat or water transport (*stiffness problem*).

2. The pressure variable has to accommodate a constant background pressure of order $\mathcal{O}_S(1)$ and the physically relevant pressure variations of order $\mathcal{O}_S(M^2)$, which leads to numerical round-off errors (*cancellation problem*).

3. For stability reasons, upwind schemes introduce artificial viscosity, which depends on the Mach number. In unfavourable settings this can cause the truncation error to grow with decreasing Mach numbers (*accuracy problem*).

The *cancellation problem* can be avoided by working only with the fluctuations of the quantities, introduced in the wave propagation approach by Leveque [34]. This method was applied to low Mach number flow by Sesterhenn *et al.* in [48]. Unfortunately, the authors do not present truly 2D-examples with shear layers – the numerical treatment of which is the heart of the *accuracy problem*.

To overcome the *stiffness and accuracy problem in steady flow simulations* a variety of time-derivative or flux preconditioning techniques have been developed and applied to compressible (and incompressible) solvers for the *inviscid flow* equations, such as Turkel's approach, [54, 55, 56], or the characteristic time stepping approach by van Leer *et al.* [58]. The preconditioning idea goes back to Chorin [9], who turned the incompressible into compressible flow equations by introducing an artificial compressibility – the resulting equations can then be solved with time advancing methods. In modern schemes the idea of manipulating the compressibility to the compressible equations is applied and generalised in order to remove or diminish the stiffness. The condition number is reduced by (almost) equalising the propagation speed of the different waves for M → 0, which accelerates the convergence to steady state. At the same time, the artificial viscosity is tuned correctly for all characteristic waves and thus the accuracy problem is circumvented. Preconditioning in the context of *viscous flow* was dealt with by Choi and Merkle in [8]. They report on the absence of the accuracy problem down to M = 10^{-6} for their preconditioning methods, which are implemented on grids with quadrilateral cells.

For atmospheric flow or flow with cavitation the transient states are of interest, so that the *stiffness problem for unsteady flow simulations* has to be dealt with. A combination of compressible flux calculation and incompressible projection methods was developed by Klein *et al.* in [27, 16] to obtain a conservative algorithm with Mach-independent accuracy and efficiency. In this class of schemes the flow variables are collocated, i.e. they are stored in the same place, the cell centres, which seems to necessitate *two projection steps*. A staggered approach, with only *one projection step* was developed by Bijl and Wesseling in [4, 57, 64].

The *accuracy problem* was explicitly addressed by Viozat and Guillard in [60, 20]. Their analysis reveals the accuracy problem of the first-order Roe scheme on *Cartesian grids*. Their numerical experiments are carried out on dual grids of triangulations. Introduced is a preconditioning of the artificial viscosity matrix, so that the stiffness in the resulting algebraic equations remains. The numerical results obtained demonstrate the successful treatment of the accuracy problem on *dual grids of triangulations*. Yet, the reduction of artificial viscosity imposes an even stronger stability constraint than the standard CFL condition for explicit time schemes as was shown by Birken *et al.* [7]. Consequently, an implicit time discretisation is obligatory for this kind of preconditioning technique. An adoption of this approach to the local Lax-Friedrichs scheme by Meister [36] corrects the wrong magnitude of the pressure field produced by the un-preconditioned scheme, but the presented pressure contour lines in the publication deviate from the solution in [60, 20].

In [19] Guillard *et al.* show that the Riemann problem itself is responsible for the creation of an unphysical pressure fluctuation of order $\mathcal{O}_S(M)$. The analysis is restricted to the one-dimensional Riemann problem and therefore does not consider two-dimensional flow situations, in which the normal velocity of leading order can be free of jumps. It will be shown in this thesis that this aspect is essential for avoiding the accuracy problem on triangular finite volume cells.

1.4. Organisation of the thesis

In this thesis it is shown, that the accuracy problem of standard first-order upwind schemes in the low Mach number regime has two facets:

1. The accuracy problem is related to the flux solver: only upwind schemes which resolve *all characteristic waves*, such as the Godunov scheme, Roe's scheme or

HLLEM do not face the accuracy problem. All other schemes like HLL, the van Leer splitting or the Steger-Warming splitting experience the accuracy problem.

2. The accuracy problem is related to the *cell geometry*: Upwind schemes like Roe's show the accuracy problem on Cartesian grids or dual grids of triangulations (comprising hexagonal cells) but not on *triangular finite volume cells*. The accuracy is maintained for a *fixed resolution* of the grid for decreasing Mach numbers.

The use of Cartesian grids or dual grids of triangulations lead to unphysical solutions for decreasing Mach numbers, see for example [60, 36]. Volpe showed in [62] for a curvilinear grid around a cylinder that a massive refinement of the quadrilateral cells is necessary to maintain the accuracy for decreasing Mach numbers.

Each facet of the accuracy problem is dedicated a part in this thesis. The relation between the accuracy problem and various flux functions is addressed in Part I, it is based on a *one-dimensional analysis*. The method of choice is the modified equation approach, which will be explained in detail in Chapter 3 with a simple transport equation. The applications to the Roe scheme and to HLL are given in Chapter 4 and 5, while Chapter 6 is dedicated to various flux vector splitting methods. All chapters close with numerical results to verify the analysis.

In Part II we analyse the relationship between the accuracy problem and the geometry of finite volume cells – it is a *two-dimensional analysis*. In the introductory Chapter 8 we present numerical results that illustrate the phenomena. An analysis for Cartesian grids is given in 9 and for triangular finite volume cells in Chapter 10. In the last chapter a rigorous proof is given, explaining why the accuracy problem does not occur on triangular finite volume cells.

To facilitate reading, some derivations of equations were put in the appendix and are referenced accordingly in the text.

2. Governing equations

The properties of a fluid can be divided into *mechanical* and *thermodynamic* properties. The equations for the mechanical properties originate in physical conservation principles. The corresponding balance laws for mass, momentum and total energy are given in the *Navier-Stokes equations* for viscid, and the *Euler equations* for inviscid flow.

Thermodynamic equations originate in the microscopic properties of a fluid and are called *equations of state*. In this thesis we restrict the analysis to *perfect gases* defined in the following section.

2.1. Equations of state for a perfect gas

A gas is called *perfect* if it satisfies a certain *thermal* and *caloric equation of state* specified in the following. References for thermodynamics and statistical physics are [15, 30, 32]. From a mathematical point of view, these equations are necessary to close the system of mechanical flow equations.

2.1.1. Thermal equation of state

The *ideal gas law* or *thermal equation of state*

$$pV = nRT, \qquad (2.1)$$

describes the relation between the state variables pressure p, volume V and absolute temperature T; n is the amount of gas and R is the universal gas constant. Fluids satisfying (2.1) are called *thermally perfect*. Equation (2.1) can be written as

$$\begin{aligned} p &= \frac{nRT}{V} \\ &= \frac{m/V}{m/n} RT \\ &= \rho \frac{R}{\mathcal{M}} T, \end{aligned}$$

with the density ρ and the molar mass $\mathcal{M} = m/n$, to obtain the form

$$\boxed{p = \rho R_{\mathrm{sp}} T} \qquad (2.2)$$

The ratio $R/\mathcal{M}$ is known as *specific gas constant* R_{sp} and depends on the gas. In all calculations presented in this thesis we used the specific gas constant for air

$$R_{\text{air}} \approx 287 \frac{\text{J}}{\text{kg} \cdot \text{K}} .$$

2.1.2. Caloric equation of state

The internal energy E_{in} of a fluid is an *extensive* property of the entire fluid considered. Relevant to computational fluid dynamics are *intensive* properties, which are defined locally, such as specific quantities or densities. If the internal energy is related to the mass of the fluid

$$e_{\text{in}} = \frac{E_{\text{in}}}{m} ,$$

we obtain the *specific internal energy* e_{in}. The internal energy per unit volume

$$\varepsilon = \frac{E_{\text{in}}}{V}$$

is also called *internal energy density*. Both intensive types of internal energy are related by

$$\varepsilon = \rho e_{\text{in}} .$$

The relation between internal specific energy e_{in} or internal specific enthalpy h_{in} on the one hand, and the absolute temperature T on the other hand, is given by the *caloric equation of state*:

$$e_{\text{in}} = c_V T , \tag{2.3a}$$
$$h_{\text{in}} = c_p T , \tag{2.3b}$$

where c_p is the *constant pressure specific heat* and c_V is the *constant volume specific heat*. The latter is given by

$$c_V = \frac{R_{\text{sp}}}{\gamma - 1} ,$$

where

$$\gamma = \frac{c_p}{c_V}$$

is the *ratio of specific heats* or *adiabatic index*. In all calculations presented in this thesis we used $\gamma = 1.4$. The constant pressure specific heat is given by

$$c_p = c_V + R_{\text{sp}} = \frac{\gamma}{\gamma - 1} R_{\text{sp}} ,$$

so that the relation between internal specific energy e_{in} and enthalpy h_{in} can be expressed as

$$h_{\text{in}} = e_{\text{in}} + \frac{p}{\rho} .$$

Equation (2.2) and (2.3) can be combined to eliminate the temperature:

$$p = \rho R_{sp} \frac{e_{in}}{c_V}$$
$$= (\gamma - 1)\rho e_{in} .$$

With the internal energy density $\varepsilon = \rho e_{in}$ the relation is simply

$$\boxed{p = (\gamma - 1)\varepsilon} \qquad (2.4)$$

The *speed of sound* in a perfect gas is given by

$$a^2 = \frac{\gamma p}{\rho} . \qquad (2.5)$$

2.1.3. The entropy

Without viscosity or other forms of dissipation and away from shocks the entropy is a conserved quantity. The Euler equations in the low Mach number regime, analysed in this thesis, have this *homentropic* property.

The relation between specific entropy s, pressure p and density ρ is

$$s = c_V \ln p - c_p \ln \rho + \text{const} ,$$

where $\ln = \log_e$ is the logarithm to the natural base e. The entropy fluctuation $\tilde{s} = s - s_0$ to a background state (p_0, ρ_0, s_0) with the corresponding entropy

$$s_0 = c_V \ln p_0 - c_p \ln \rho_0 + \text{const}$$

is given by

$$\boxed{\tilde{s} = c_V \ln \frac{p}{p_0} - c_p \ln \frac{\rho}{\rho_0}} \qquad (2.6)$$

In homentropic flow we have $\tilde{s} = 0$, so that we obtain

$$c_p \ln \frac{\rho}{\rho_0} = c_V \ln \frac{p}{p_0} ,$$

which can be written as

$$\left(\frac{\rho}{\rho_0}\right)^\gamma = \frac{p}{p_0} ,$$

or simply as

$$\boxed{p \sim \rho^\gamma} \qquad (2.7)$$

We point out here already, that a numerical effect called *artificial viscosity* has similar properties compared to the *physical viscosity*, among the similarities is the production of entropy. Therefore the analysis of entropy plays a major role in this thesis for understanding the behaviour of numerical schemes.

2.2. The Navier-Stokes equations

The Navier-Stokes equations describe the mechanical behaviour of a viscous flow and are the conservation principles for mass, momentum and total energy applied to a fluid. In *dimensional* form they read

$$\frac{\partial}{\partial \hat{t}}\hat{\rho} + \hat{\nabla} \cdot (\hat{\rho}\hat{\mathbf{u}}) = 0 \,,$$

$$\frac{\partial}{\partial \hat{t}}\hat{\rho}\hat{\mathbf{u}} + \hat{\nabla} \cdot ((\hat{\rho}\hat{\mathbf{u}}) \circ \hat{\mathbf{u}}) + \hat{\nabla}\hat{p} = \hat{\nabla} \cdot \hat{\Pi} \,,$$

$$\frac{\partial}{\partial \hat{t}}\hat{\rho}\hat{e} + \hat{\nabla} \cdot (\hat{\mathbf{u}}(\hat{\rho}\hat{e} + \hat{p})) = \hat{\nabla} \cdot (\hat{\Pi}\hat{\mathbf{u}}) \,,$$

where the *dimensional* quantities are the density $\hat{\rho}$, the momentum density $\hat{\rho}\hat{\mathbf{u}}$, the pressure $\hat{p}$ and the total energy per unit volume $\hat{\rho}\hat{e}$. The stress tensor $\hat{\Pi}$ is given by

$$\hat{\Pi} = \eta(\hat{\nabla}\hat{\mathbf{u}} + (\hat{\nabla}\hat{\mathbf{u}})^T - \frac{2}{3}(\hat{\nabla} \cdot \hat{\mathbf{u}})D) \,,$$

where η is the coefficient of dynamic viscosity and $D = \text{diag}(1,1,1)$. The nabla operator with a hat $\hat{\nabla}$ symbolises the dimensional derivative in space. These evolution equations describe a compressible flow of a viscous fluid of Newtonian type without heat conduction or external forces. Note that all physical variables depend on the space variable $\mathbf{x} \in \mathbb{R}^3$ and on the time $t \in \mathbb{R}_0^+$. For ease of reading we do not write this information explicitly whenever it is not stated otherwise.

For later analysis we are interested in the behaviour of the primitive variables $\hat{\rho}$, $\hat{\mathbf{u}}$ and $\hat{p}$ given by the following set of equations:

$$\frac{\partial}{\partial \hat{t}}\hat{\rho} + \hat{\nabla} \cdot (\hat{\rho}\hat{\mathbf{u}}) = 0 \,,$$

$$\frac{\partial}{\partial \hat{t}}\hat{\mathbf{u}} + \hat{\mathbf{u}} \cdot \hat{\nabla}\hat{\mathbf{u}} + \frac{1}{\hat{\rho}}\hat{\nabla}\hat{p} = \frac{1}{\hat{\rho}}\hat{\nabla} \cdot \hat{\Pi} \,,$$

$$\frac{\partial}{\partial \hat{t}}\hat{p} + \hat{\nabla} \cdot (\hat{\mathbf{u}}\hat{p}) + (\gamma - 1)\hat{p}\hat{\nabla} \cdot \hat{\mathbf{u}} = (\gamma - 1)\hat{\Pi} : \hat{\nabla}\hat{\mathbf{u}} \,.$$

The symbol ":" is a double scalar product as defined in e.g. [5].

Non-dimensionalisation

Low Mach number flow can be investigated using asymptotic analysis, which requires all variables to be free of physical units and to be of the same order of magnitude. For this purpose we replace all dimensional variables $\hat{\phi}$ by the product of their dimensional scaling or reference quantity ϕ_{ref} and a dimensionless variable ϕ:

$$\hat{\phi} \rightsquigarrow \phi_{\text{ref}} \cdot \phi \,.$$

The following quantities and their reference quantities will be relevant in our study:

ρ_{ref} the reference of the density
u_{ref} the reference of the flow velocity
l_{ref} and t_{ref} the reference of length and time scale
p_{ref} the reference of the pressure, the internal energy $\hat{e}$
                and the total energy $\hat{\rho}\hat{e}$

The last scalings make sense because $\hat{p} = (\gamma - 1)\hat{e}$ for perfect gases and $\hat{\rho}\hat{e} \approx \hat{e}$ for low Mach number flow. p_{ref} is also the reference quantity of the total enthalpy per unit volume $\hat{\rho}\hat{h} = \hat{\rho}\hat{e} + \hat{p}$. For scaling the speed of sound we set

$$a_{\text{ref}}^2 = p_{\text{ref}}/\rho_{\text{ref}}$$

and obtain for the sound speed in perfect gases in dimensionless form:

$$a^2 = \gamma p/\rho \, .$$

If we scale an independent variable like $\hat{t}$ by a reference quantity t_{ref}, the corresponding differential operator is transformed in the following way:

$$\hat{t} = t\, t_{\text{ref}} \quad \Rightarrow \quad \frac{\mathrm{d}}{\mathrm{d}\hat{t}} = \frac{1}{t_{\text{ref}}} \frac{\mathrm{d}}{\mathrm{d}t} \, . \tag{2.8}$$

Introducing these reference quantities into the compressible Navier-Stokes equations we obtain:

$$\text{Str}\, \rho_t + \nabla \cdot (\rho \mathbf{u}) = 0 \, ,$$
$$\text{Str}\, \mathbf{u}_t + \mathbf{u} \cdot \nabla \mathbf{u} + \frac{1}{\text{M}^2}\frac{1}{\rho}\nabla p = \frac{1}{\rho \text{Re}} \nabla \cdot \Pi \, , \tag{2.9}$$
$$\text{Str}\, p_t + \mathbf{u} \cdot \nabla p + \gamma p \nabla \cdot \mathbf{u} = \frac{\text{M}^2}{\text{Re}}(\gamma - 1)\Pi : \nabla \mathbf{u} \, .$$

The Mach number

$$\text{M} = \frac{u_{\text{ref}}}{a_{\text{ref}}}$$

is the ratio of magnitudes of flow velocity and sound speed. The Strouhal number

$$\text{Str} = \frac{l_{\text{ref}}/t_{\text{ref}}}{u_{\text{ref}}}$$

relates the scales of space and time with the fluid velocity and the Reynolds number

$$\text{Re} = \frac{l_{\text{ref}}\rho_{\text{ref}}u_{\text{ref}}}{\eta}$$

relates convective to viscous transport. Let us now examine the limit equations for $\text{Re} \to 0$ and $\text{Re} \to \infty$.

Stokes flow

If the transport by dissipation exceeds the convective transport by orders of magnitude, the Navier-Stokes equations are transformed into a diffusion dominated system. The resulting flow type – called *creeping flow* or *Stokes flow* – is characterised by a vanishing Reynolds numbers
$$\text{Re} \to 0.$$
The smallness of this flow parameter can be exploited to build asymptotic expansions of the physical quantities, which will elucidate the physics behind *creeping flow*. To this end let us first introduce the related mathematical concepts.

2.2.1. Introduction to asymptotic analysis

Landau symbols

Following [25, 22] we assume two scalar functions $u(\mathbf{x}; \epsilon)$ and $v(\mathbf{x}; \epsilon)$ to be dependent on the space variable $\mathbf{x}$ and a scalar parameter ϵ.

Definition 2.2.1. *The statement*
$$u(\mathbf{x}; \epsilon) = \mathcal{O}(v(\mathbf{x}; \epsilon)) \quad as \quad \epsilon \to 0$$
*on some bounded domain Ω is equivalent to saying that **u is bounded by v** in the following sense: for all $\mathbf{x}$ in Ω there is a positive number $k(\mathbf{x})$, such that*
$$|u(\mathbf{x}; \epsilon)| \leq k(\mathbf{x}) |v(\mathbf{x}; \epsilon)| . \tag{2.10}$$
*If k is independent of $\mathbf{x}$ this property is called **uniformly valid on** Ω.*

Definition 2.2.2. *If we can interchange the functions u and v so that in addition to (2.10) $v = \mathcal{O}(u)$ we write*
$$v(\mathbf{x}; \epsilon) = \mathcal{O}_S(u(\mathbf{x}; \epsilon)) \quad as \quad \epsilon \to 0 ,$$
*in which case we say that **u and v are of the same order of magnitude on** Ω.*

Definition 2.2.3. *Two functions u and v are called **asymptotically equal**, written as*
$$u(\mathbf{x}; \epsilon) \sim v(\mathbf{x}; \epsilon) \quad as \quad \epsilon \to 0 ,$$
if
$$\lim_{\epsilon \to 0} \frac{u(\mathbf{x}; \epsilon)}{v(\mathbf{x}; \epsilon)} = 1 .$$

Definition 2.2.4. *The term **u is arbitrarily small compared to v** is written as*
$$u(\mathbf{x}; \epsilon) = o(v(\mathbf{x}; \epsilon)) \quad as \quad \epsilon \to 0$$
and means that for all $\mathbf{x}$ in Ω and $\delta > 0$ there is an ϵ-interval $0 < \epsilon < \epsilon_1$ so that
$$|u(\mathbf{x}; \epsilon)| \leq \delta |v(\mathbf{x}; \epsilon)| .$$

Presence of two small parameters In the analysis we often use the $\mathcal{O}_S$-symbol in conjunction with two small parameters, the Mach number M and the grid size Δx. We want to understand a term such as $\mathcal{O}_S(\Delta x M)$ in the following sense:

$$\mathcal{O}_S(\Delta x M) = \begin{cases} \mathcal{O}_S(\Delta x) & \text{as } \Delta x \to 0, \\ \mathcal{O}_S(M) & \text{as } M \to 0. \end{cases}$$

If not stated otherwise, we assume a fixed grid resolution $\Delta x = \text{const}$ and use M as asymptotic parameter. The term $\mathcal{O}_S(\Delta x M)$ only underlines the dependency of numerical errors on the grid size. We do not investigate consistency in this thesis, i.e. the behaviour for $\Delta x \to 0$, which we assume given for the schemes we investigate.

Asymptotic sequence and asymptotic expansion

A sequence of real functions $\phi_n(\epsilon)$ is called *asymptotic sequence*, if for $n = 1, 2, \dots$

$$\phi_{n+1} = o(\phi_n(\epsilon)) \quad \text{as} \quad \epsilon \to 0.$$

In what follows we will need the power series $\phi_n = \epsilon^{n-1}$ of the parameter ϵ, which is an asymptotic sequence, since

$$\lim_{\epsilon \to 0} \frac{\phi_{n+1}(\epsilon)}{\phi_n(\epsilon)} = \lim_{\epsilon \to 0} \epsilon = 0.$$

We want to write $u(\mathbf{x}; \epsilon)$ in terms of functions $u_n(\mathbf{x})$ depending only on space and an asymptotic sequence $\phi_n(\epsilon)$ depending only on the parameter ϵ. The series

$$\sum_{n=1}^{N} \phi_n(\epsilon) u_n(\mathbf{x})$$

is called *asymptotic $(N+1)$-term expansion* of $u(\mathbf{x}; \epsilon)$ as $\epsilon \to 0$ with respect to the sequence $\phi_n(\epsilon)$ if

$$u(\mathbf{x}; \epsilon) - \sum_{n=1}^{N} \phi_n(\epsilon) u^{(n)}(\mathbf{x}) = o(\phi_N) \quad \text{as} \quad \epsilon \to 0. \tag{2.11}$$

If the function $u(\mathbf{x}; \epsilon)$ is given along with a sequence $\phi_n(\epsilon)$, we can apply Equation (2.11) to find the uniquely defined asymptotic terms of $u(\mathbf{x}; \epsilon)$:

$$u^{(1)}(\mathbf{x}) = \lim_{\epsilon \to 0} \frac{u(\mathbf{x}; \epsilon)}{\phi_1(\epsilon)},$$

$$u^{(2)}(\mathbf{x}) = \lim_{\epsilon \to 0} \frac{u(\mathbf{x}; \epsilon) - \phi_1(\epsilon) u_1(\mathbf{x})}{\phi_2(\epsilon)},$$

$$\vdots$$

$$u^{(k)}(\mathbf{x}) = \lim_{\epsilon \to 0} \frac{u(\mathbf{x}; \epsilon) - \sum_{n=1}^{k-1} \phi_n(\epsilon) u_n(\mathbf{x})}{\phi_k(\epsilon)}.$$

Note that, unlike the original function u, the asymptotic functions $u^{(k)}$ do no longer depend on the small parameter ϵ. Having underlined this fact once, we do not explicitly mention it in the following asymptotic analyses given in this thesis.

2.2.2. Asymptotic analysis for small Reynolds numbers

The special case of viscous flow with very low Reynolds numbers Re is of interest, since interesting parallels to numerical effects can be drawn. The momentum equation for incompressible viscous flow is given by

$$\text{Str}\frac{\partial}{\partial t}\mathbf{u} + \mathbf{u}\cdot\nabla\mathbf{u} + \frac{1}{M^2}\frac{1}{\rho}\nabla p = \frac{1}{\text{Re}}\frac{1}{\rho}\nabla\cdot\Pi\,, \tag{2.12}$$

where we can drop the time derivatives since we are interested in steady flow only. Furthermore, we express the Reynolds number Re in terms of the Mach number M to avoid the problem with *two* small parameters described in [28]. For this reason we write

$$\text{Re} = \frac{\rho_{\text{ref}}l_{\text{ref}}}{\eta}a_{\text{ref}}\frac{u_{\text{ref}}}{a_{\text{ref}}} = \chi M\,, \tag{2.13}$$

where

$$\chi = \frac{\rho_{\text{ref}}l_{\text{ref}}a_{\text{ref}}}{\eta}$$

is a dimensionless quantity. If we decrease the Mach number solely by lowering the velocity u_{ref} of the flow, while keeping the speed of sound and all other quantities constant, χ is also constant. As a consequence, the Reynolds number is proportional to the Mach number and can be expressed as in (2.13). For the steady Navier-Stokes equation we then have

$$\mathbf{u}\cdot\nabla\mathbf{u} + \frac{1}{M^2}\frac{1}{\rho}\nabla p = \frac{1}{\chi M}\frac{1}{\rho}\nabla\cdot\Pi\,.$$

Let us assume the existence of an asymptotic expansion for the pressure p, the density ρ and the velocity $\mathbf{u}$ for $M \to 0$. Introducing the asymptotic 3-term expansions

$$\rho = \rho^{(0)} + M\rho^{(1)} + M^2\rho^{(2)} + o(M^2)\,,$$
$$u = u^{(0)} + Mu^{(1)} + M^2u^{(2)} + o(M^2)\,,$$
$$p = p^{(0)} + Mp^{(1)} + M^2p^{(2)} + o(M^2)\,,$$

into (2.12) and separating the equation according to powers of M, we obtain an equivalent system of asymptotic equations, cf. [36]. They are in detail, sorted by powers of M:

order M^{-2}:
$$\nabla p^{(0)} = 0\,, \tag{2.14}$$

order M^{-1}:
$$\nabla p^{(1)} = \frac{1}{\chi}\nabla\cdot\Pi^{(0)}\,, \tag{2.15}$$

leading order M^0:
$$u^{(0)} \cdot \nabla u^{(0)} + \frac{\nabla p^{(2)}}{\rho^{(0)}} = \frac{1}{\chi} \frac{\nabla \cdot \Pi^{(1)}}{\rho^{(0)}} \,. \tag{2.16}$$

Note that $\Pi^{(0)}$ and $\Pi^{(1)}$ are asymptotic functions of Π, which are defined via the asymptotic functions $\mathbf{u}^{(i)}$, $i = 0, 1, 2$. An asymptotic expansion in this way is well defined, cf. [36]. The divergence constraint for the leading order velocity

$$\nabla \cdot \mathbf{u}^{(0)} = 0 \tag{2.17}$$

can be derived from the (constant density) continuity equation. The asymptotic equations allow for the following interpretations:

- The leading order pressure given in (2.14) is a spatially constant background pressure.

- The pressure variations in the flow, given in (2.15), are of order $\mathcal{O}_S(M)$. Recall that in inviscid low Mach number flow it is of order $\mathcal{O}_S(M^2)$. Note that we use the symbol $\mathcal{O}_S$ (to be of the same order) instead of $\mathcal{O}$ (bounded by) to underline that the pressure fluctuations are indeed *of same order*. This is legitimate if we exclude trivial flows without any pressure fluctuations.

- The pressure gradient $\nabla p^{(1)}$ balances the viscosity term in (2.15) and not the transport term in the leading-order momentum Equation (2.16).

2.2.3. Stokes flow around a cylinder

As a test case for our numerical schemes we consider the flow around a cylinder and look for analytical solutions in the case of excessive viscosity. C. G. Stokes was the first to solve the flow around a sphere in a highly viscous medium. He assumed small Reynolds numbers

$$\mathrm{Re} = \frac{\rho_\infty u_\infty R}{\eta} \,,$$

which can be achieved with a small radius R of the sphere, a sufficiently small free-stream velocity u_∞ or a large coefficient of dynamic viscosity η. His idea was to drop all transport terms and external force terms in the momentum equation and look for solutions to the system

$$\nabla p = \eta \nabla^2 \mathbf{u} \,, \tag{2.18}$$

$$\nabla \cdot \mathbf{u} = 0 \,, \tag{2.19}$$

with no-slip boundary conditions at the surface of the sphere, cf. [31]. Comparing these equations (with their asymptotic counterparts (2.15) and (2.17), the following terms have to be identified with each other:

$$p^{(1)} \rightsquigarrow p, \quad \mathbf{u}^{(0)} \rightsquigarrow \mathbf{u}, \quad \rho^{(0)} \rightsquigarrow \rho \,.$$

The background pressure $p^{(0)}$ has to be identified with the pressure at infinity

$$p^{(0)} \rightsquigarrow p_\infty ,$$

it has no dynamical function – we only need it for defining a reference Mach number later on. The physical flow domain is $\mathbb{R}^3$. Although System (2.18) can be solved for the sphere, it can be shown that there is no solution to these equations for the problem in the plane [29], i.e. for the flow around a cylinder. Oseen showed in [43] that keeping a linearised form of the transport terms leads to a system of equations

$$u_\infty \nabla \mathbf{u} + \frac{1}{\rho}\nabla p = \frac{1}{\rho}\nabla^2 \mathbf{u} ,$$

$$\nabla \cdot \mathbf{u} = 0 ,$$

which can be solved analytically, where u_∞ is the free-stream velocity. The solution to this equation system is given in [29] in terms of radial velocity u_r and azimuthal velocity u_θ:

$$u_r = \frac{A_0}{r} - \frac{A_1 \cos\theta}{r^2} + u_\infty \cos\theta$$
$$- C_0 \left\{ \frac{1}{2kr} + \frac{1}{2}\cos\theta - \frac{1}{2}\cos\theta \ln\left(\frac{1}{2}\gamma k r\right) \right\} + \frac{C_1 \cos\theta}{2kr^2} ,$$

$$u_\theta = \frac{A_1 \sin\theta}{r^2} - u_\infty \sin\theta - C_0 \frac{\sin\theta}{2} \ln\left(\frac{1}{2}\gamma k r\right) + \frac{C_1 \sin\theta}{2kr^2} ,$$

and for the pressure

$$p = p_\infty - \rho u_\infty \frac{\partial \phi}{\partial x} ,$$

where ϕ is given by

$$\phi = A_0 \ln r + A_1 \frac{\partial \ln r}{\partial x} + A_2 \frac{\partial^2 \ln r}{\partial x^2} + \ldots ,$$

and A_0, A_1, A_2, C_0, C_1 are constants that have to be specified using the boundary conditions. The no-slip boundary condition of the Navier-Stokes equation at the cylinder surface $r = R$:

$$u_r(R) = 0 ,$$
$$u_\theta(R) = 0 ,$$

leads to a pressure field approximation valid near the surface of the cylinder:

$$\boxed{p = p_\infty - \rho u_\infty A_0 \frac{\cos\theta}{r}}$$

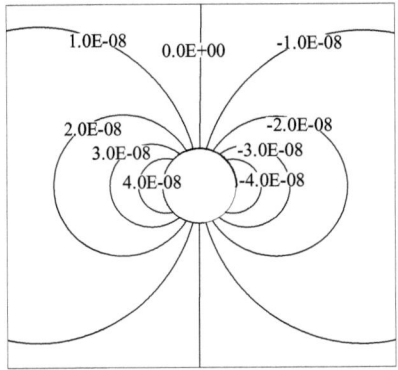

Figure 2.1.: Isolines of the pressure fluctuation $\tilde{p} = (p - p_\infty)/p_\infty$ for the creeping flow around a cylinder for a free-stream Mach number $M_\infty = 10^{-3}$.

where A_0 is a constant depending on the Reynolds number and the viscosity, cf. [29] for details. The isolines of the pressure fluctuation

$$\tilde{p} = \frac{p - p_\infty}{p_\infty}$$

for a free-stream Mach number $M_\infty = 10^{-3}$ are given in Figure 2.1. Note that in an asymptotic sense the Mach number is still finite and the speed of sound is given by the background values of pressure and density:

$$a_\infty^2 = \frac{\gamma p_\infty}{\rho},$$

leading to a well-defined free-stream Mach number $M_\infty = u_\infty/a_\infty$. We come back to this result in Chapter 5, where we analyse the pressure field obtained with the HLL scheme for low Mach numbers. Interestingly, it looks very similar to this approximate solution of the creeping flow equations around a cylinder.

2.3. The Euler equations

In a variety of flows, e. g. of a fluid with a small viscosity coefficient, with a large flow velocity or large length scales, the Reynolds number is large. In the limit

$$\text{Re} \to \infty$$

the viscous terms in the Navier-Stokes equation can be neglected and we obtain the *Euler equations*:

$$\operatorname{Str} \rho_t + \nabla \cdot (\rho \mathbf{u}) = 0 \,,$$
$$\operatorname{Str} \mathbf{u}_t + \mathbf{u} \cdot \nabla \mathbf{u} + \frac{1}{M^2}\frac{1}{\rho}\nabla p = 0 \,, \qquad (2.20)$$
$$\operatorname{Str} p_t + \mathbf{u} \cdot \nabla p + \gamma p \nabla \cdot \mathbf{u} = 0 \,,$$

which describe a compressible flow without dissipation of momentum or energy. Note that due to viscosity boundary layers develop on surfaces, in which flow variables can change on a small length scale, so that the local Reynolds number is not large $\operatorname{Re}_{local} \ll \infty$. Nevertheless, on a large scale or away from boundary layers, the Euler equations can be an accurate model.

For the further analysis we look for a relation between the Strouhal number Str and the Mach number M.

2.3.1. The Strouhal number

The meaning of the dimensionless number

$$\operatorname{Str} = \frac{l_{\mathrm{ref}}}{u_{\mathrm{ref}} t_{\mathrm{ref}}} = \frac{l_{\mathrm{ref}} f_{\mathrm{ref}}}{u_{\mathrm{ref}}} \qquad (2.21)$$

was first established by Strouhal in 1878. He compared a characteristic frequency f_{ref} of an unsteady, periodic flow phenomenon and a typical length scale l_{ref} with the velocity v_{ref} of the flow. Typical examples of the appearance of this dimensionless number are vibrations of aerofoils or the frequency of the shedding of eddies in a flow behind a body as illustrated by J. Zierep in [66, 67]. We use a more general notion of this dimensionless number starting from the Euler equations. The term

$$\frac{l_{\mathrm{ref}}/t_{\mathrm{ref}}}{u_{\mathrm{ref}}}$$

appears as factor of the time derivative in the Euler equations after scaling each equation by the reference convection $u_{\mathrm{ref}}/l_{\mathrm{ref}}$. The factor $1/t_{\mathrm{ref}}$ of the time derivative is a measure of the (reciprocal) time scale. The Strouhal number therefore reflects the proportion of convection on a given time scale: on the acoustic time scale we obtain the transport of the physical quantities by sound waves, while on the flow time scale we obtain the transport by convection.

Flow phenomena

If we focus on the flow, the time scale is given by the time an average fluid particle with an average velocity u_{ref} needs to pass the flow domain of length l_{ref}:

$$t_{\mathrm{ref}} = \frac{l_{\mathrm{ref}}}{u_{\mathrm{ref}}} \,,$$

which, substituted in (2.21), gives the Strouhal number

$$\boxed{\text{Str}_{\text{flow}} = 1}$$

Acoustic phenomena

With focus on acoustic phenomena, the time scale is now linked to the speed of sound

$$t_{\text{ref}} = \frac{l_{\text{ref}}}{a_{\text{ref}}},$$

so that (2.21) gives a different Strouhal number:

$$\boxed{\text{Str}_{\text{ac}} = \frac{1}{M}}$$

In this case the Strouhal number introduces the inverse Mach number as factor of the time derivatives in the Euler equations. We will later see in the asymptotic analysis that the equations with Str = 1/M describe the fluid dynamics in terms of acoustic wave behaviour.

So far, only the *ratio* of space and time scale was fixed, namely to the speed of sound, but not the scales themselves. Fixing them leads to different situations:

Short wave acoustics: If we choose our reference length l_{ref} to be the scale of the flow domain l_{flow}, the asymptotic equations describe short wave acoustics. The time scale of this type of acoustics

$$t_{\text{ref}} = \frac{l_{\text{ref}}}{\text{Str}\, u_{\text{ref}}} = M\, t_{\text{flow}}$$

is $\mathcal{O}_S(M)$ compared to the time scale of the flow phenomena. For small Mach numbers the change of physical quantities due to acoustics is so quick, that there is no interaction with the flow. This decoupling of sound is very familiar to us, e.g. if we think of the undistorted transport of sound in an orchestra room that always has some amount of thermal convection.

Long wave acoustics: An essential interaction between acoustics and flow can be expected if both phenomena exist on the same time scale t_{ref}. In this case the length scales of flow l_{flow} and acoustics $l_{\text{ac}} = l_{\text{ref}}$ have to differ by orders of magnitude according to

$$l_{\text{ref}} = \text{Str}\, u_{\text{ref}} t_{\text{ref}} = \frac{1}{M} l_{\text{flow}}\, .$$

The breeze of air in front of a sub-woofer is a flow generated by long wave acoustics. A profound analysis of the interaction of long scale acoustics and fluid flow necessitates a multi-scale asymptotics, which was first done by Klein in [27].

Strouhal number in numerical computations

In explicit schemes, the choice of the time step $\Delta\hat{t}$ for a given cell size $\Delta\hat{x}$ is restricted by the Courant-Friedrichs-Levy (CFL) condition, stating that the grid speed $\Delta\hat{x}/\Delta\hat{t}$ should be less than the fastest signal speed, which is approximately the speed of sound in low Mach number flow. We conclude from the (unscaled) CFL condition

$$\frac{\Delta\hat{x}}{\Delta\hat{t}} = \mathcal{O}_S(\hat{a}) \,. \tag{2.22}$$

If we scale grid size and time step with the given length and time scale

$$\Delta\hat{x} = l_{\text{ref}}\Delta x \,,$$
$$\Delta\hat{t} = t_{\text{ref}}\Delta t \,,$$

Equation (2.22) reads:

$$\frac{\Delta\hat{x}}{\Delta\hat{t}} = \frac{l_{\text{ref}}}{t_{\text{ref}}}\frac{\Delta x}{\Delta t} = \mathcal{O}_S(a_{\text{ref}}a) \,.$$

From the order relation of the CFL condition in nondimensional form

$$\frac{\Delta x}{\Delta t} = \mathcal{O}_S(a) \,,$$

we can extract an order relation for the CFL condition of the reference quantities:

$$\frac{l_{\text{ref}}}{t_{\text{ref}}} = \mathcal{O}_S(a_{\text{ref}}) \,. \tag{2.23}$$

As a consequence the Strouhal number in explicit schemes is

$$\boxed{\text{Str}_{\text{num}} = \frac{1}{M}}$$

which has implications for the asymptotic analysis of upwind schemes with an explicit time discretisation in the low Mach number regime.

2.3.2. Asymptotic analysis for small Mach numbers

An order analysis of the physical quantities interacting with each other in low Mach number flow motivates an asymptotic analysis of the Euler equations for low Mach numbers.

Physical motivation

We consider the total energy density of the flow

$$\hat{\rho}\hat{e} = \hat{\varepsilon} + \frac{1}{2}\hat{\rho}\hat{u}^2 \,,$$

where $\hat{u}$ is the absolute value of the dimensional fluid velocity $\hat{\mathbf{u}}$ and $\hat{\varepsilon}$ the internal energy density. We scale this equation with the reference of total energy $e_{\text{ref}} = p_{\text{ref}}$ and obtain its non-dimensional form

$$\rho e = \varepsilon + \frac{1}{2}\rho u^2 \mathrm{M}^2 \;.$$

The kinetic energy is $\mathcal{O}_{\mathrm{S}}(\mathrm{M}^2)$ compared to the internal energy. No matter how slow the fluid moves, if it encounters an obstacle like an aerofoil the fluid is locally accelerated. By energy conservation (Bernoulli's principle, cf. [61]), the change of pressure due to acceleration can only be $\mathcal{O}_{\mathrm{S}}(\mathrm{M}^2)$ like the kinetic energy. In steady flow a splitting of the pressure

$$p = p^{(0)} + \mathrm{M}^2 p^{(2)}$$

is therefore physically motivated.

Let us assume an unsteady flow where any pressure perturbations are transported via acoustic waves. Which order terms of pressure p and velocity u interact in these waves? To find an answer, we start with the energy density of an acoustic wave (see for example [61])

$$\hat{e}_{\text{pot}} = \frac{1}{2}\hat{\kappa}(\Delta \hat{p})^2, \tag{2.24}$$

with $\hat{\kappa}$ being the compressibility of the fluid, which is

$$\hat{\kappa} = \frac{1}{\hat{a}^2 \hat{\rho}}$$

for perfect gases. Scaling Equation (2.24) with p_{ref} will not introduce any dimensionless numbers since $a_{\text{ref}}^2 = p_{\text{ref}}/\rho_{\text{ref}}$. We replace the pressure fluctuation Δp in the non-dimensional version of (2.24) by $p^{(n)}\mathrm{M}^n$, with $n \in \mathbb{N}$ being the order of the pressure perturbation. The order of magnitude of the potential energy density in the acoustic wave is then

$$\hat{e}_{\text{pot}} = \frac{1}{2}\frac{(\Delta \hat{p})^2}{\hat{a}^2 \hat{\rho}} \quad \Rightarrow \quad e_{\text{pot}} = \frac{1}{2}\frac{(p^{(n)}\mathrm{M}^n)^2}{a^2 \rho} = \mathcal{O}_{\mathrm{S}}(\mathrm{M}^{2n}) \;.$$

The kinetic energy density

$$\hat{e}_{\text{kin}} = \frac{1}{2}\hat{\rho}\hat{u}_{\mathrm{S}}^2 \;,$$

with $\hat{u}_{\mathrm{S}}$ representing the (dimensional) particle velocity due to the acoustic wave (German: *Schallschnelle*), can again be scaled with the reference pressure p_{ref} and the reference flow velocity u_{ref} to obtain

$$e_{\text{kin}} = \frac{1}{2}\rho u_{\mathrm{S}}^2 \mathrm{M}^2 \;.$$

This kinetic energy density must be $\mathcal{O}(\mathrm{M}^{2n})$ like the potential energy density, which forces the particle velocity $\mathbf{u}_{\mathrm{S}}$ in the acoustic wave to be $\mathcal{O}_{\mathrm{S}}(\mathrm{M}^{n-1})$:

$$\mathbf{u}_{\mathrm{S}} = \mathrm{M}^{n-1}\mathbf{u}^{(n-1)} \;.$$

For example, a pressure perturbation $p^{(1)}$ is linked to a velocity perturbation $\mathbf{u}^{(0)}$ in an acoustic wave, therefore we call them *asymptotic couple*. In this way $p^{(2)}$ and $\mathbf{u}^{(1)}$ also form an acoustic couple. The splitting of the fluid velocity into a background flow velocity of leading order and the compression-related particle velocities of higher orders

$$\mathbf{u} = \mathbf{u}^{(0)} + M\mathbf{u}^{(1)} + M^2\mathbf{u}^{(2)} + \ldots ,$$

is therefore physically founded.

Time scale of the flow

We now turn to the mathematical analysis of the Euler equations on the time scale of the flow. Note that in this context

$$\mathrm{Str}_{\mathrm{flow}} = 1 .$$

Let us assume the existence of a asymptotic 3-term expansion of the physical quantities:

$$\begin{aligned}
\rho &= \rho^{(0)} + M\rho^{(1)} + M^2\rho^{(2)} + o(M^2) \quad \text{as} \quad M \to 0 , \\
p &= p^{(0)} + Mp^{(1)} + M^2p^{(2)} + o(M^2) \quad \text{as} \quad M \to 0 , \\
\mathbf{u} &= \mathbf{u}^{(0)} + M\mathbf{u}^{(1)} + M^2\mathbf{u}^{(2)} + o(M^2) \quad \text{as} \quad M \to 0 ,
\end{aligned} \quad (2.25)$$

on a bounded domain Ω. Inserting the asymptotic expansion (2.25) into the Equations (2.20) and separating them with respect to powers of the Mach number [36, 27], we obtain the following enlarged system of equations:

$$\rho_t^{(0)} + \nabla \cdot (\rho\mathbf{u})^{(0)} = 0 , \qquad (2.26\mathrm{a})$$

$$\rho_t^{(1)} + \nabla \cdot (\rho\mathbf{u})^{(1)} = 0 , \qquad (2.26\mathrm{b})$$

$$\rho_t^{(2)} + \nabla \cdot (\rho\mathbf{u})^{(2)} = 0 , \qquad (2.26\mathrm{c})$$

$$\nabla p^{(0)} = 0 , \qquad (2.26\mathrm{d})$$

$$\nabla p^{(1)} = 0 , \qquad (2.26\mathrm{e})$$

$$\mathbf{u}_t^{(0)} + \mathbf{u}^{(0)} \cdot \nabla \mathbf{u}^{(0)} + \left(\frac{\nabla p}{\rho}\right)^{(2)} = 0 , \qquad (2.26\mathrm{f})$$

$$p_t^{(0)} + \mathbf{u}^{(0)} \cdot \nabla p^{(0)} + \gamma p^{(0)} \nabla \cdot \mathbf{u}^{(0)} = 0 , \qquad (2.26\mathrm{g})$$

$$p_t^{(1)} + (\mathbf{u} \cdot \nabla p)^{(1)} + \gamma (p\nabla \cdot \mathbf{u})^{(1)} = 0 , \qquad (2.26\mathrm{h})$$

$$p_t^{(2)} + (\mathbf{u} \cdot \nabla p)^{(2)} + \gamma (p\nabla \cdot \mathbf{u})^{(2)} = 0 . \qquad (2.26\mathrm{i})$$

Expressions in brackets with a superscript (n) define the entire term to be of that order and its expansion follows the rules given in Appendix A.3.

The analysis and interpretation of this set of equations can be found for example in [26, 27]. Only the essential results will be summarised here: the pressure of leading and first order are constant in space:

$$p^{(0)} = p^{(0)}(t)$$

$$p^{(1)} = p^{(1)}(t)$$

and its gradients can be omitted in the Equations (2.26d) to (2.26i). Equation (2.26g) for the leading order pressure can be written as

$$\frac{p_t^{(0)}}{\gamma p^{(0)}} = -\nabla \cdot \mathbf{u}^{(0)} \tag{2.27}$$

or after integrating over the whole bounded domain Ω

$$\frac{1}{\gamma p^{(0)}} \frac{\mathrm{d} p^{(0)}}{\mathrm{d} t} = -\frac{1}{|\Omega|} \oint \mathbf{u}^{(0)} \cdot \mathbf{n} \, \mathrm{d} A \tag{2.28}$$

Equation (2.28) implies that a global compression increases the background pressure $p^{(0)}$ and vice versa. On the other hand, Equation (2.27) implies that an increase in background pressure leads to a divergence $\nabla \cdot \mathbf{u}^{(0)}$ – everywhere and at the same time in the fluid. It is worth recalling, that this statement is true on the time scale of the flow for $\mathrm{M} \to 0$. For a temporally constant background pressure, $p^{(0)} = \mathrm{const}$, Equation (2.27) turns into the classical divergence constraint

$$\nabla \cdot \mathbf{u}^{(0)} = 0$$

for incompressible flow.

The second order pressure $p^{(2)}$ functions as the balancing force of this incompressible flow field

$$\mathbf{u}_t^{(0)} + \mathbf{u}^{(0)} \cdot \nabla \mathbf{u}^{(0)} + \frac{1}{\rho^{(0)}} \nabla p^{(2)} = 0 \ .$$

The acoustic time scale

Let us now turn to the asymptotic analysis of the Euler equations on the acoustic time sale. Setting time and space scale according to $l_\mathrm{ref}/t_\mathrm{ref} = a_\mathrm{ref}$ leads to the Strouhal number:

$$\mathrm{Str} = \frac{a_\mathrm{ref}}{u_\mathrm{ref}} = \frac{1}{\mathrm{M}} \ .$$

With this choice of temporal perspective, the asymptotic equations describe phenomena on the acoustic time scale.

We substitute Str $= 1/\mathrm{M}$ into the nondimensional Euler equations (2.20) to obtain

$$\rho_t + \mathrm{M}\,\nabla \cdot (\rho \mathbf{u}) = 0\,,$$
$$\mathbf{u}_t + \mathrm{M}\,\mathbf{u}\cdot\nabla\mathbf{u} + \frac{1}{\mathrm{M}}\frac{1}{\rho}\nabla p = 0\,, \qquad (2.29)$$
$$p_t + \mathrm{M}\,\mathbf{u}\cdot\nabla p + \mathrm{M}\,\gamma p \nabla\cdot\mathbf{u} = 0\,,$$

and replace the physical variables by their asymptotic expansion (2.25) to obtain the asymptotic system of equations:

$$\rho_t^{(0)} = 0\,, \qquad (2.30\mathrm{a})$$
$$\rho_t^{(1)} + \nabla \cdot (\rho^{(0)}\mathbf{u}^{(0)}) = 0\,, \qquad (2.30\mathrm{b})$$
$$\rho_t^{(2)} + \nabla \cdot (\rho\mathbf{u})^{(1)} = 0\,, \qquad (2.30\mathrm{c})$$

$$\frac{1}{\rho^{(0)}}\nabla p^{(0)} = 0\,, \qquad (2.30\mathrm{d})$$
$$\mathbf{u}_t^{(0)} + \left(\frac{\nabla p}{\rho}\right)^{(1)} = 0\,, \qquad (2.30\mathrm{e})$$
$$\mathbf{u}_t^{(1)} + \mathbf{u}^{(0)}\cdot\nabla\mathbf{u}^{(0)} + \left(\frac{\nabla p}{\rho}\right)^{(2)} = 0\,, \qquad (2.30\mathrm{f})$$
$$\mathbf{u}_t^{(2)} + (\mathbf{u}\cdot\nabla\mathbf{u})^{(1)} = 0\,, \qquad (2.30\mathrm{g})$$

$$p_t^{(0)} = 0\,, \qquad (2.30\mathrm{h})$$
$$p_t^{(1)} + \mathbf{u}^{(0)}\cdot\nabla p^{(0)} + \gamma p^{(0)}\nabla\cdot\mathbf{u}^{(0)} = 0\,, \qquad (2.30\mathrm{i})$$
$$p_t^{(2)} + (\mathbf{u}\cdot\nabla p)^{(1)} + \gamma(p\nabla\cdot\mathbf{u})^{(1)} = 0\,. \qquad (2.30\mathrm{j})$$

Equation (2.30a) implies that the leading order density is constant in time

$$\boxed{\rho^{(0)} = \rho^{(0)}(x)}$$

and therefore, to leading order, the material transport due to convection of the flow is frozen on this time scale. From (2.30d) and (2.30h) we conclude that the background pressure is constant in space *and* time

$$\boxed{p^{(0)} = \mathrm{const}}$$

A global compression

$$\oint \mathbf{u}^{(0)} \cdot \mathbf{n} \, dA > 0$$

does no longer lead to an instantaneous increase of the leading order pressure $p^{(0)}$, as on the time scale of the flow, but to pressure perturbations in $p^{(1)}$.

Wave equation for $p^{(1)}$ and $\mathbf{u}^{(0)}$ Simplifying Equations (2.30e) and (2.30i) using the constancy of $p^{(0)}$, we obtain

$$\mathbf{u}_t^{(0)} + \frac{1}{\rho^{(0)}} \nabla p^{(1)} = 0 , \tag{2.31}$$

$$p_t^{(1)} + \gamma p^{(0)} \nabla \cdot \mathbf{u}^{(0)} = 0 , \tag{2.32}$$

which illustrates that $p^{(1)}$ and $\mathbf{u}^{(0)}$ are an *acoustic couple*: the gradient of $p^{(1)}$ is the accelerating agent for $\mathbf{u}^{(0)}$ and the compression $\nabla \cdot \mathbf{u}^{(0)}$ leads to a pressure increase of $p^{(1)}$.

If we take the divergence of (2.31) and the time derivative of (2.32), we can eliminate $\nabla \cdot \mathbf{u}_t^{(0)}$ to obtain for $p^{(1)}$:

$$p_{tt}^{(1)} - [a^{(0)}]^2 \nabla^2 p^{(1)} = -\frac{[a^{(0)}]^2}{\rho^{(0)}} \nabla \rho^{(0)} \cdot \nabla p^{(1)} .$$

In case of constant $\rho^{(0)}$ this is a homogeneous wave equation for $p^{(1)}$ with the phase velocity $[a^{(0)}]^2 = \gamma p^{(0)}/\rho^{(0)}$. In the absence of external compression

$$p^{(1)} = \text{const} ,$$

once any initial perturbations in $p^{(1)}$ have left the flow domain in *external flows*. In *internal flows* (with perfectly reflecting boundaries) perturbations can, theoretically, live forever as there is no dissipation inside the fluid. With the time derivative of (2.31) and the gradient of (2.32), we can deduce a wave equation for $\mathbf{u}^{(0)}$

$$\mathbf{u}_{tt}^{(0)} - [a^{(0)}]^2 \nabla(\nabla \cdot \mathbf{u}^{(0)}) = 0$$

with the same phase velocity $[a^{(0)}]^2 = \gamma p^{(0)}/\rho^{(0)}$.

Wave equation for $p^{(2)}$ and $\mathbf{u}^{(1)}$ In the evolution equation for $p^{(2)}$, (2.30j), with all asymptotic terms expanded:

$$p_t^{(2)} = -\mathbf{u}^{(0)} \cdot \nabla p^{(1)} + \gamma(p^{(0)} \nabla \cdot \mathbf{u}^{(1)} + p^{(1)} \nabla \cdot \mathbf{u}^{(0)}) , \tag{2.33}$$

the gradient $\nabla p^{(1)}$ and the divergence $\nabla \cdot \mathbf{u}^{(0)}$ vanish if we assume no acoustic waves of this order. The remaining part

$$p_t^{(2)} = \gamma p^{(0)} \nabla \cdot \mathbf{u}^{(1)}$$

couples pressure changes in $p^{(2)}$ with a first order compression or divergence $\nabla \cdot \mathbf{u}^{(1)}$. Expanding Equation (2.30f) and dropping $\nabla p^{(0)}$ and $\nabla p^{(1)}$ we obtain

$$\mathbf{u}_t^{(1)} + \mathbf{u}^{(0)} \cdot \nabla \mathbf{u}^{(0)} + \frac{1}{\rho^{(0)}} \nabla p^{(2)} = 0,$$

where the gradient of $p^{(2)}$ acts as an accelerating agent for the first-order velocity. Therefore $\mathbf{u}^{(1)}$ and $p^{(2)}$ form a further *acoustic couple*. The resulting wave equation

$$p_{tt}^{(2)} - [a^{(0)}]^2 \nabla^2 p^{(2)} = \gamma p^{(0)} (\nabla(\mathbf{u}^{(0)} \cdot \nabla \mathbf{u}^{(0)})) \tag{2.34}$$

has a source term that is related to the Lighthill tensor [35], which is used to describe the sound generation by vortices. The wave equation for $\mathbf{u}^{(1)}$ can be obtained in a similar way:

$$\mathbf{u}_{tt}^{(1)} - [a^{(0)}]^2 \nabla(\nabla \cdot \mathbf{u}^{(1)}) = -(\mathbf{u}^{(0)} \cdot \nabla \mathbf{u}^{(0)})_t . \tag{2.35}$$

2.3.3. Incompressible potential flow around a cylinder

Let us now turn to a further special type of flow, which can be derived from the Euler equations for low Mach numbers: the incompressible potential flow. Recall, with the assumption of constant background pressure

$$p^{(0)} = \text{const},$$

the leading-order velocity has the divergence constraint

$$\nabla \cdot \mathbf{u}^{(0)} = 0 .$$

Under the additional assumption of zero rotation

$$\nabla \times \mathbf{u}^{(0)} = 0,$$

we can write $\mathbf{u}^{(0)}$ as a gradient of a potential $\Phi^{(0)}$

$$\mathbf{u}^{(0)} = \nabla \Phi^{(0)} .$$

To agree with standard notations in the literature, we identify from now on for the dynamical quantities:

$$\Phi^{(0)} \rightsquigarrow \Phi, \quad \mathbf{u}^{(0)} \rightsquigarrow \mathbf{u}, \quad p^{(2)} \rightsquigarrow p .$$

For the thermodynamic background state we identify

$$p^{(0)} \rightsquigarrow p_\infty, \quad \rho^{(0)} \rightsquigarrow \rho,$$

which are needed for defining a reference Mach number M_∞. The divergence constraint expressed with Φ is the potential equation

$$\boxed{\nabla^2 \Phi = 0}$$

Constant density irrotational flow can therefore be solved analytically using the tools of potential theory. The analytical solution to the *incompressible potential flow around a cylinder* serves as reference solution for many numerical experiments in this thesis. Its potential is given (cf. [49]) in polar coordinates

$$\Phi = u_\infty \left(r + \frac{R^2}{r} \right) \cos \phi ,$$

with the cylinder radius R and the velocity at infinity u_∞. For the radial component u_r we obtain

$$u_r = \frac{\partial \Phi}{\partial r} = u_\infty (1 - \alpha) \cos \phi, \quad \text{where} \quad \alpha := \frac{R^2}{r^2}$$

and for the angular velocity component

$$u_\phi = \frac{1}{r} \frac{\partial \Phi}{\partial \phi} = -u_\infty (1 + \alpha) \sin \phi .$$

To derive the pressure p with Bernoulli's principle for incompressible flow, we need the kinetic energy. The absolute value $u := \|\mathbf{u}\|$ of the velocity is given by

$$\begin{aligned} u^2 &= u_r^2 + u_\phi^2 = u_\infty^2 \{ (1-\alpha)^2 \cos^2 \phi + (1+\alpha)^2 \sin^2 \phi \} \\ &= u_\infty^2 \{ 1 - 2\alpha \cos 2\phi + \alpha^2 \} . \end{aligned}$$

Bernoulli's law

$$p + \frac{1}{2} \rho u^2 = \text{const}$$

with the reference state $(\rho, u_\infty, p_\infty)$ at infinity gives

$$\begin{aligned} p &= p_\infty + \frac{\rho}{2} (u_\infty^2 - u^2) \\ &= p_\infty + \frac{\rho}{2} u_\infty^2 (2\alpha \cos 2\phi - \alpha^2) . \end{aligned}$$

Let us define the *(scaled) pressure fluctuation* $\tilde{p}$ as

$$\tilde{p} := \frac{p - p_\infty}{p_\infty} .$$

For the potential flow around the cylinder $\tilde{p}$ we obtain

$$\boxed{\tilde{p}(\alpha, \phi) = \frac{\gamma}{2} \mathrm{M}_\infty^2 (2\alpha \cos 2\phi - \alpha^2)} \qquad (2.36)$$

where we have defined the Mach number at infinity $\mathrm{M}_\infty = u_\infty / a_\infty$ with the sound speed for perfect gases corresponding to the background state:

$$a_\infty^2 = \frac{\gamma p_\infty}{\rho} .$$

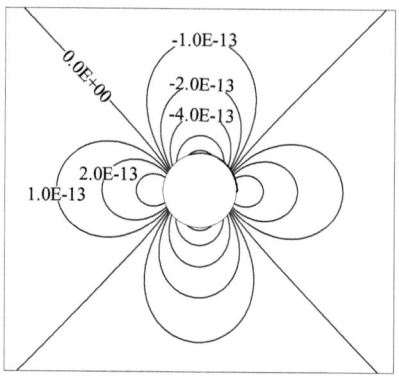

Figure 2.2.: Isovalues of the pressure fluctuation $\tilde{p}$ for the incompressible potential flow around a cylinder for a free-stream Mach number $M_\infty = 10^{-6}$.

Note that the sound speed in incompressible media with constant density is, formally, infinity. To allow comparison with numerical data of low Mach number flows, we have to define the sound speed in the described way in an asymptotic context. The pressure fluctuation at the stagnation point ($\phi = 0$, $r = R$) has a global maximum of

$$\tilde{p}_{\max} = \tilde{p}(\phi = 0, r = R) = \frac{\gamma}{2}M_\infty^2 \;.$$

For symmetry reasons this pressure is also valid at ($\phi = \pi$, $r = R$). At the upper and lower point of the cylinder, with coordinates ($\phi = \pm\pi/2$, $r = R$), the pressure fluctuation takes the global minimum

$$\tilde{p}_{\min} = \tilde{p}(\phi = \pm\pi/2, r = R) = -\frac{3}{2}\gamma M_\infty^2 \;,$$

so that the maximum pressure fluctuation p_{fluc} in the flow field is given by

$$p_{\text{fluc}} = p_{\max} - p_{\min} = 2\gamma M_\infty^2 \;. \tag{2.37}$$

The contour lines of the pressure fluctuation are given in Figure 2.2 for a free-stream Mach number of $M_\infty = 10^{-6}$.

Part I.
Analysis for 1D: flux solver

3. Concept of numerical Reynolds number

In this part we investigate the behaviour of first-order upwind schemes in a *one-dimensional model situation*. It is useful to distinguish between two classes of upwind schemes: those which resolve all characteristic waves, *complete Riemann solvers*, and those which resolve only parts of the characteristic spectrum, *incomplete Riemann solvers*. This *completeness* property is an important aspect for understanding the *accuracy problem* of upwind schemes in the low Mach number regime. It can be analysed with the concepts of artifical viscosity and numerical Reynolds number in conjunction with the modified equation approach.

3.1. Introduction

Numerical experiments on grids with triangular finite volume cells, structured as well as unstructured, suggest that the behaviour of first-order upwind schemes in the low Mach number regime differs enormously. The reason for the dependency on the grid

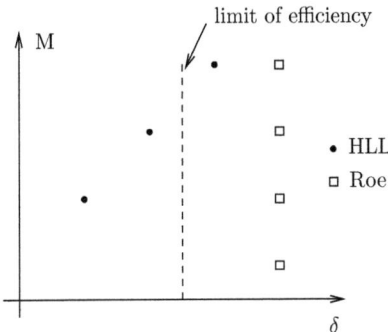

Figure 3.1.: Necessary refinement in terms of cell size δ to simulate flows with decreasing Mach numbers to a given accuracy. Box symbols: first-order Roe scheme. Circle symbols: first-order HLL.

cell geometry can only be understood with a two-dimensional analysis as presented in Part II.

Schemes like HLL, on the one hand, show the so called *accuracy problem*: the errors produced by the scheme grow with decreasing Mach numbers on a given grid. As depicted schematically in Figure 3.1, for a Mach number of M ≈ 0.1 the scheme produces accurate results on a given grid with cell size δ_0. To maintain this accuracy for smaller Mach numbers refinement is necessary. This numerical behaviour was first described by Volpe in [62].

The Roe scheme, on the other hand, does not show the *accuracy problem*: it maintains the quality on the same grid for decreasing Mach numbers. As it turned out, the different behaviour can be understood with the concept of artificial viscosity.

3.2. Physical and artificial viscosity

Recall, the Reynolds number Re compares viscous to convective transport of momentum:

$$\text{Re} = \frac{\text{magnitude of momentum convection}}{\text{magnitude of momentum dissipation}}$$

where the magnitude of convection and dissipation are given by the reference quantities in the flow:

$$\text{Re} = \left\{\frac{\rho_\text{ref} v_\text{ref}^2}{l_\text{ref}}\right\} / \left\{\frac{\eta v_\text{ref}}{l_\text{ref}^2}\right\} = \frac{\rho_\text{ref} v_\text{ref} l_\text{ref}}{\eta} \ .$$

A similar number can be defined for a numerical scheme, although the viscosity terms originate in a numerical truncation error. It will give us an order relation for the numerical effects we want to keep small, compared to the physical effects we investigate.

3.2.1. The modified equation approach

One possible way to understand the behaviour of upwind schemes in the low Mach number regime is the analysis of the *modified equation*. We introduce this concept with the linear advection equation

$$q_t + a q_x = 0 \ , \tag{3.1}$$

where a is the constant velocity of advection and q the transported quantity. The upwind scheme, also called Courant-Isaacson-Rees method (CIR), for this equation with $a < 0$ is given by

$$\frac{q_i^{n+1} - q_i^n}{\Delta t} = -\frac{a}{\Delta x}(q_{i+1}^n - q_i^n) \ , \tag{3.2}$$

where q_i^n approximates the solution of (3.1) in cell i at the time t_n. We replace in (3.2) the discrete values q_i by a smooth solution $q(x,t)$ of the original PDE

$$\frac{q(x_i, t_n + \Delta t) - q(x_i, t_n)}{\Delta t} = -a \frac{q(x_i + \Delta x, t_n) - q(x_i, t_n)}{\Delta x} \ ,$$

and expand the terms in Taylor series about (x_i, t_n)

$$\frac{q_t \Delta t + \frac{1}{2} q_{tt} \Delta t^2 + \mathcal{O}(\Delta t^3)}{\Delta t} = -a \frac{q_x \Delta x + \frac{1}{2} q_{xx} \Delta x^2 + \mathcal{O}(\Delta x^3)}{\Delta x} .$$

Keeping only the lowest-order terms in Δx and Δt, we obtain a so called *modified differential equation* of the discrete equation (3.2) of the upwind method:

$$q_t + a q_x = -\frac{1}{2} q_{tt} \Delta t - \frac{1}{2} a q_{xx} \Delta x . \tag{3.3}$$

The terms on the RHS are called truncation error. They are for the temporal and spatial discretisation of first order, in agreement with the classification of the CIR as a *first-order scheme*.

An equation of the same accuracy as (3.3) without the temporal derivative q_{tt} can be derived using a method as given in [63]. For this purpose we derive from (3.3) the relations

$$q_t \doteq -a q_x ,$$

$$q_{xt} \doteq -a q_{xx} ,$$

$$q_{tt} \doteq -a q_{xt} \doteq a^2 q_{xx} ,$$

where $\doteq$ means equal except for first order terms (cf. [13]), and substitute them into (3.3) to obtain another form of the modified equation:

$$q_t + a q_x = -\frac{1}{2} a^2 q_{xx} \Delta t - \frac{1}{2} a q_{xx} \Delta x . \tag{3.4}$$

The derivation of the modified equation for $a > 0$

$$q_t + a q_x = -\frac{1}{2} a^2 q_{xx} \Delta t + \frac{1}{2} a q_{xx} \Delta x \tag{3.5}$$

can be found in [34]. For an arbitrary sign of the advection velocity a we combine (3.4) and (3.5) to obtain another form of the *modified equation* of the CIR method:

$$\boxed{q_t + a q_x = \frac{1}{2} |a| q_{xx} \Delta x (1 - c) = \frac{1}{2} \eta_{\text{num}} q_{xx}} \tag{3.6}$$

where $c = |a| \Delta t / \Delta x$ is the Courant number. Compared to the original partial differential equation (PDE), Equation (3.6) has an additional second derivative with respect to x, which has a dissipative effect on the solution, cf. [68]. The term

$$\eta_{\text{num}} = |a| \Delta x (1 - c)$$

is therefore called *coefficient of artificial or numerical viscosity*; it is a measure of the magnitude of artificial dissipation.

Derivatives of order higher than two are only of secondary importance for first-order schemes we focus on in this thesis. The situation changes if we consider 2nd-order schemes like Fromm-, Lax-Wendroff- or the Beam-Warming method. Then a spatial derivative of 3rd order constitutes the leading order deviation from the original partial differential equation, which manifests itself by dispersive effects, i.e. oscillations due to different group velocities [52, 53].

3.2.2. Reynolds numbers for numerical schemes

Even if the viscosity terms originate in a numerical method, we can define a number that relates the magnitude of convective transport to the magnitude of artificial dissipation and call it *numerical Reynolds number*:

$$\text{Re}_{\text{num}} = \frac{\text{magnitude of convection}}{\text{magnitude of artificial dissipation}}$$

Since the numerical Reynolds number contains the term $1/\Delta x$, its limit for vanishing grid size is

$$\lim_{\Delta x \to 0} \text{Re}_{\text{num}} = \infty \, .$$

This expresses the consistency of the numerical scheme with the inviscid model equation. Recall that the inviscid Euler equations can be seen as limit equations for viscous flow with $\text{Re} \to \infty$. We want to point out that

1. in low Mach number flows the magnitude of convection decreases with the Mach number and

2. for practical calculations the grid size δ is always finite, maintaining a certain amount of artificial viscosity.

The main difference between upwind schemes will turn out to be the dependency of the artificial viscosity on the Mach number.

3.3. Upwind methods for linear scalar equations

We want to apply the concept of the numerical Reynolds number to variants of the upwind method for the linear advection equation. How do different first-order spatial and temporal discretisation methods influence the dissipative behaviour of the scheme?

3.3.1. Explicit upwind method (CIR)

In the last section we found the modified equation for the CIR method:

$$q_t + a q_x = \frac{1}{2}|a|\Delta x(1 - c)q_{xx} \, .$$

In order to find the magnitude of convection and dissipation we introduce the following reference quantities that are characteristic for the magnitude of the physical quantities in the system:

q_{ref} for the transported quantity,

l_{ref} for the size of the investigated flow domain,

a_{ref} for the advection speed,

t_{ref} for the time scale.

Note that the time scale is linked to length scale and characteristic advection speed by

$$a_{\text{ref}} = \frac{l_{\text{ref}}}{t_{\text{ref}}} .$$

All scaled quantities are then $\mathcal{O}(1)$, except Δx, which of course is $\mathcal{O}_S(\Delta x)$. The magnitude of physical convection is given by

$$a_{\text{ref}} \frac{q_{\text{ref}}}{l_{\text{ref}}} ,$$

and the magnitude of artificial dissipation by

$$a_{\text{ref}} \frac{q_{\text{ref}}}{l_{\text{ref}}^2} (1-c) l_{\text{ref}} \Delta x .$$

The numerical Reynolds number for the CIR method is therefore

$$\boxed{\text{Re}_{\text{CIR}} = \frac{1}{\Delta x (1-c)}}$$

The inviscid advection equation, characterised by $\text{Re}_{\text{num}} \to \infty$, is obtained for

a) $\Delta x \to 0$ (consistency),

b) $c \to 1$ (exact upwinding).

Note that exact upwind methods do not exist for the nonlinear Euler equations.

3.3.2. Implicit upwind method

We consider the upwind scheme with the implicit Euler time discretisation. For a positive advection speed $a > 0$ it is given by

$$q_i^{n+1} = q_i^n - \frac{a \Delta t}{\Delta x} (q_i^{n+1} - q_{i-1}^{n+1}) ,$$

or by

$$q(x_i, t_{n+1}) = q(x_i, t_n) - \frac{a \Delta t}{\Delta x} \left[q(x_i, t_{n+1}) - q(x_{i-1}, t_{n+1}) \right] , \quad (3.7)$$

with a smooth solution $q(x,t)$ of the original PDE. We expand the first term on the RHS of (3.7) in a Taylor series about (x_i, t_{n+1}) using $t_n = t_{n+1} - \Delta t$:

$$q(x_i, t_{n+1} - \Delta t) = q(x_i, t_{n+1}) - q_t(x_i, t_{n+1}) \Delta t + \frac{1}{2} q_{tt}(x_i, t_{n+1}) \Delta t^2 + \ldots . \quad (3.8)$$

Inserting (3.8) into Equation (3.7) and dropping all higher order terms we obtain

$$q_t + a q_x = \frac{1}{2} q_{tt} \Delta t + \frac{1}{2} a q_{xx} \Delta x . \quad (3.9)$$

We want to express all time derivatives on the right hand side by space derivatives. Using (3.9) we find the first-order relations

$$q_{xt} \doteq -aq_{xx},$$
$$q_{tt} \doteq -aq_{tx} \doteq a^2 q_{xx},$$

and inserting them in (3.9) gives

$$q_t + aq_x = \frac{1}{2}a^2 q_{xx}\Delta t + \frac{1}{2}aq_{xx}\Delta x.$$

For the implicit CIR method with $a < 0$ we can derive

$$q_t + aq_x = \frac{1}{2}a^2 q_{xx}\Delta t - \frac{1}{2}aq_{xx}\Delta x,$$

so that we obtain for the modified equation for a general a:

$$q_t + aq_x = \frac{1}{2}|a|\Delta x(1+c)q_{xx}, \quad \text{with} \quad c = |a|\frac{\Delta t}{\Delta x}.$$

The numerical Reynolds number can be found as

$$\boxed{\text{Re}_{\text{CIR,im}} = \frac{1}{\Delta x(1+c)} \sim \frac{1}{\Delta x}}$$

The different dependency on the Courant number c, compared to the CIR method, shows that for a given grid the implicit upwind method is more diffusive.

3.3.3. Implicit upwind method with large time steps

In the Euler equations the acoustic and entropy waves travel at velocities that differ by orders of magnitude in the low Mach number regime. To avoid a break down of the convergence rate to the steady solution, large time steps according to the time scale of the flow are necessary, which can be achieved with implicit time discretisation methods to avoid the restrictive CFL condition. But this implies large Courant numbers for the acoustic waves, i. e. the acoustic waves travel across many cells during a single time step. We can model this situation with an implicit scheme for the linear advection equation making the following assumptions: the advection wave travels at an average speed of a_{ref} but the time step

$$t_{\text{ref}} := \frac{l_{\text{ref}}}{u_{\text{ref}}}$$

is chosen according to a slower velocity u_{ref} of a fictitious wave, which will be the entropy wave in the Euler equations. Also in analogy to the Euler equations, we introduce a parameter

$$\text{M} = \frac{u_{\text{ref}}}{a_{\text{ref}}}$$

as the ratio of the two wave speeds. Since M is a measure of the stiffness of the equation we also refer to it as *stiffness parameter*. Note that a small stiffness parameter corresponds to a large stiffness. If we allow large time steps of order $\mathcal{O}_S(1/M)$, the Courant number for the fast waves will be $c = \mathcal{O}_S(1/M)$. The new scaling of the Courant number causes the numerical Reynolds number to be of a different order:

$$\boxed{\text{Re}_{\text{num}} = \frac{1}{\Delta x (1+c)} \sim \frac{M}{\Delta x} \quad \text{as} \quad M \to 0}$$

This asymptotic equality can be interpreted in the following way:

- For a given grid, i.e. with fixed Δx, a stiffness parameter $M \to 0$ leads to a vanishing numerical Reynolds number $\text{Re}_{\text{num}} \to 0$, or put in other words: the artificial dissipation term exceeds the physical advection term by orders of magnitude. As a consequence the modified equation changes its type from hyperbolic to parabolic and the physical nature of the advection equation gets completely lost.

- In order to approximate inviscid flow ($\text{Re}_{\text{num}} \to \infty$) for a small parameter M, the grid has to be extremely refined, i.e. $\Delta x = o(M)$, a phenomenon that was reported by Volpe in [62] in the context of the low Mach number Euler equations.

3.4. Summary

As already known for stiff ordinary differential equations [12], consistency is not a sufficient concept to describe the quality of a numerical scheme. For the low Mach number Euler equations consistency is neither sufficient to describe the different behaviour of upwind schmes. We therefore propose a new concept which makes statements on the usefulness of upwind schemes for a *fixed finite grid cell size* in the case of decreasing Mach numbers:

Definition 3.4.1. *A first-order upwind scheme for the linear advection equation is called **asymptotically consistent** with respect to a stiffness parameter M, if the corresponding modified equation of second-order accuracy has a numerical Reynolds number satisfying*

$$\text{Re}_{\text{num}} = \mathcal{O}_S\left(\frac{1}{\Delta x}\right) \quad \text{as} \quad M \to 0 \,.$$

This order relation depends only on the reciprocal cell size Δx and not on the Mach number. If the numerical Reynolds number also depends on M in the form

$$\text{Re}_{\text{num}} = \mathcal{O}_S\left(\frac{M}{\Delta x}\right) \quad \text{as} \quad M \to 0 \,,$$

*the scheme is called **asymptotically inconsistent** with respect to the (small) stiffness parameter M.*

Theorem 3.4.2. *The following numerical schemes for the linear advection equation are **asymptotically consistent** with respect to the stiffness parameter* M:

- *the CIR method (cf. Section 3.3.1)*
- *the implicit upwind method with small time steps, i. e. with Courant numbers of order $\mathcal{O}_S(1)$ (cf. Section 3.3.2)*

*The following numerical schemes for the linear advection equation are **asymptotically inconsistent** with respect to the stiffness parameter* M:

- *the implicit upwind scheme with large time steps, i. e. with Courant numbers of order $\mathcal{O}_S(1/\mathrm{M})$, (cf. Section 3.3.3)*

Proof. The proof for the diverse variants of the upwind scheme for the linear advection equation can be found in the corresponding sections. □

4. The Roe scheme

In this chapter we analyse the asymptotic behaviour of numerical schemes using the approximate Riemann solver by P. L. Roe [45] with different time discretisation methods. For this purpose the modified equation in characteristic variables are derived, to see the viscosity effects on each individual characteristic wave. This necessitates the reduction of the two-dimensional flow equations to a one-dimensional model situation.

4.1. Characteristic equations

The 2D-Euler equations in conservation-law form are

$$\mathbf{q}_t + \mathbf{f}(\mathbf{q})_x + \mathbf{g}(\mathbf{q})_y = 0 \ ,$$

with

$$\mathbf{q} = \begin{bmatrix} \rho \\ \rho u \\ \rho v \\ \rho e \end{bmatrix}, \quad \mathbf{f} = \begin{bmatrix} \rho u \\ \rho u^2 + p \\ \rho u v \\ u(\rho e + p) \end{bmatrix}, \quad \mathbf{g} = \begin{bmatrix} \rho v \\ \rho u v \\ \rho v^2 + p \\ v(\rho e + p) \end{bmatrix}.$$

They can generally not be transformed into a decoupled system with characteristic variables due to the different eigenspaces of $A = \mathrm{d}\mathbf{f}/\mathrm{d}\mathbf{q}$ and $B = \mathrm{d}\mathbf{g}/\mathrm{d}\mathbf{q}$, cf. [34]. In this chapter we therefore assume only plane waves – parallel to the y-axis without loss of generality – so that the 2D-equations collapse to a one-dimensional problem given by

$$\mathbf{q}_t + \mathbf{f}(\mathbf{q})_x = 0 \ . \tag{4.1}$$

Note, that $\mathbf{q}$ and $\mathbf{f}$ still depend on *two* velocity components: the normal velocity u and the transverse velocity v, so that shear waves are possible. In the following chapters we analyse this equation on the real line and refer to it as the *one-dimensional model case*.

System (4.1) can be (locally) transformed to characteristic form using the right eigenvectors $\mathbf{r}_i$ of the Jacobian $A = \mathrm{d}\mathbf{f}/\mathrm{d}\mathbf{q}$ with the transformation matrix

$$R = [\mathbf{r}_1, \ldots, \mathbf{r}_4] \tag{4.2}$$

and its inverse

$$R^{-1} = L = \begin{pmatrix} \mathbf{l}_1^T \\ \vdots \\ \mathbf{l}_4^T \end{pmatrix} \tag{4.3}$$

to obtain
$$\frac{\partial \mathbf{w}}{\partial t} + \Lambda \frac{\partial \mathbf{w}}{\partial x} = 0,$$
where the characteristic variables are given by

$$dw_1 = \frac{1}{2a^2}\,dp - \frac{\rho}{2a}\,du, \quad \text{(left-running acoustic wave)}$$

$$dw_2 = d\rho - \frac{dp}{a^2}, \quad \text{(entropy wave)}$$

$$dw_3 = \rho\,dv, \quad \text{(shear wave)}$$

$$dw_4 = \frac{1}{2a^2}\,dp + \frac{\rho}{2a}\,du. \quad \text{(right-running acoustic wave)}$$

4.2. The first-order Roe scheme

The numerical dissipation of the explicit Roe scheme is twofold: with only spatial derivative discretised, i. e. with the semi-discrete form, we can find the dissipative effects of the approximate Riemann solver. The fully-discrete form shows in addition the dissipative effects of the temporal discretisation.

4.2.1. Spatial discretisation

The Roe scheme [45] applied to the *one-dimensional model case* (4.1) can be written in semi-discrete form as

$$\frac{\mathrm{d}}{\mathrm{d}t}\mathbf{q}_i = -\frac{1}{\Delta x}(\mathbf{f}^{\text{Roe}}_{i+1/2} - \mathbf{f}^{\text{Roe}}_{i-1/2}), \tag{4.4}$$

with the numerical flux function

$$\mathbf{f}_{\text{Roe}}(\mathbf{q}_R, \mathbf{q}_L) = \frac{\mathbf{f}(\mathbf{q}_R) + \mathbf{f}(\mathbf{q}_L)}{2} - \frac{1}{2}|\hat{A}(\mathbf{q}_R, \mathbf{q}_L)|(\mathbf{q}_R - \mathbf{q}_L), \tag{4.5}$$

where $\hat{A}$ denotes the Roe matrix for the right state $\mathbf{q}_R$ and the left state $\mathbf{q}_L$ of the Riemann problem. Inserting (4.5) into (4.4), we obtain the equation of the semi-discrete Roe scheme:

$$\frac{\mathrm{d}}{\mathrm{d}t}\mathbf{q}_i = -\frac{\mathbf{f}_{i+1} - \mathbf{f}_{i-1}}{2\Delta x} + \frac{1}{2}\frac{|\hat{A}_{i+1,i}|(\mathbf{q}_{i+1} - \mathbf{q}_i) - |\hat{A}_{i,i-1}|(\mathbf{q}_i - \mathbf{q}_{i-1})}{\Delta x}. \tag{4.6}$$

Let $\mathbf{q}(x,t)$ be a smooth solution to the original PDE and $\mathbf{f}$ the corresponding smooth flux function. For the first expression on the RHS of (4.6) we obtain

$$\frac{\mathbf{f}_{i+1} - \mathbf{f}_{i-1}}{2\Delta x} = \mathbf{f}_x\big|_{x_i} + \mathcal{O}(\Delta x^2) = A(\mathbf{q})\mathbf{q}_x\big|_{x_i} + \mathcal{O}(\Delta x^2),$$

which is a second-order accurate approximation of the physical flux derivative. For the second term on the RHS of (4.6) we need an expansion of the Roe matrix. For this purpose we interpret the Roe average $\mathbf{q}_{i+1,i}$ as a deviation $\Delta \mathbf{q}$ of the state $\mathbf{q}_i$

$$\mathbf{q}_{i+1,i} = \mathbf{q}_i + \Delta \mathbf{q} ,$$

where the Roe average is defined as

$$\mathbf{q}_{i+1,i} = \theta \mathbf{q}_i + (1-\theta)\mathbf{q}_{i+1} , \quad \text{with} \quad \theta = \frac{\sqrt{\rho_i}}{\sqrt{\rho_i} + \sqrt{\rho_{i+1}}} .$$

This can be written in terms of $\mathbf{q}(x,t)$ to obtain

$$\begin{aligned}
\mathbf{q}_{i+1,i} &= \theta \mathbf{q}(x_i) + (1-\theta)\mathbf{q}(x_i + \Delta x) \\
&= \theta \mathbf{q}(x_i) + (1-\theta)[(\mathbf{q}(x_i) + \mathbf{q}_x(x_i)\Delta x + \mathcal{O}(\Delta x^2)] \\
&= \mathbf{q}(x_i) + (1-\theta)[(\mathbf{q}_x(x_i)\Delta x + \mathcal{O}(\Delta x^2)] .
\end{aligned}$$

The weight θ itself is a function of $\rho(x)$ and can be expanded in a Taylor series about $\rho(x_i)$:

$$\theta(\rho(x_i + \Delta x)) = \theta(\rho(x_i)) + \mathcal{O}(\Delta x) = \frac{1}{2} + \mathcal{O}(\Delta x) . \tag{4.7}$$

Using (4.7) we obtain for the Roe average

$$\mathbf{q}_{i+1,i} = \mathbf{q}(x_i) + \left(\frac{1}{2} + \mathcal{O}(\Delta x)\right)\left(\mathbf{q}_x(x_i)\Delta x + \mathcal{O}(\Delta x^2)\right) \tag{4.8}$$

$$= \mathbf{q}(x_i) + \frac{1}{2}\mathbf{q}_x(x_i)\Delta x + \mathcal{O}(\Delta x^2) , \tag{4.9}$$

so that the deviation $\Delta \mathbf{q}$ between Roe average and inner cell state is

$$\Delta \mathbf{q} = \frac{1}{2}\mathbf{q}_x(x_i)\Delta x .$$

The Roe matrix expanded in a Taylor series about $\mathbf{q}_i = \mathbf{q}(x_i)$ is therefore

$$\begin{aligned}
\hat{A}_{i+1,i} &= A(\mathbf{q}_i) + \frac{\partial A}{\partial \mathbf{q}}\Delta \mathbf{q} + \mathcal{O}(\Delta \mathbf{q}^2) \\
&= A(\mathbf{q}_i) + \frac{\partial A}{\partial \mathbf{q}}\frac{1}{2}\mathbf{q}_x\Delta x + \mathcal{O}(\Delta \mathbf{q}^2) \\
&= A(\mathbf{q}_i) + \mathcal{O}(\Delta x) ,
\end{aligned}$$

which gives

$$|\hat{A}_{i+1,i}| = |A(\mathbf{q}_i)| + \mathcal{O}(\Delta x) \tag{4.10}$$

for the absolute value of the Roe matrix. We can now substitute (4.8) and (4.10) into the viscosity term

$$\eta_{\text{Roe}} := \frac{|\hat{A}_{i+1,i}|(\mathbf{q}_{i+1} - \mathbf{q}_i) - |\hat{A}_{i,i-1}|(\mathbf{q}_i - \mathbf{q}_{i-1})}{2\Delta x},$$

to obtain

$$\begin{aligned}\eta_{\text{Roe}} &= \frac{1}{2}\left(|A(\mathbf{q}_i)| + \mathcal{O}(\Delta x)\right)\left(\mathbf{q}_x\big|_{x_{i+1/2}} + \mathcal{O}(\Delta x^2)\right) \\ &\quad - \frac{1}{2}\left(|A(\mathbf{q}_i)| + \mathcal{O}(\Delta x)\right)\left(\mathbf{q}_x\big|_{x_{i-1/2}} + \mathcal{O}(\Delta x^2)\right) \\ &= \frac{1}{2}|A(\mathbf{q}_i)|\left(\mathbf{q}_x\big|_{x_{i+1/2}} - \mathbf{q}_x\big|_{x_{i-1/2}}\right) + \mathcal{O}(\Delta x)\frac{1}{2}\left(\mathbf{q}_x\big|_{x_{i+1/2}} - \mathbf{q}_x\big|_{x_{i-1/2}}\right) \\ &= \frac{1}{2}|A(\mathbf{q}_i)|\,\mathbf{q}_{xx}\big|_{x_i}\,\Delta x + \mathcal{O}(\Delta x^2).\end{aligned}$$

We can summarise the results in the following modified equation:

$$\boxed{\mathbf{q}_t + A(\mathbf{q})\mathbf{q}_x = \frac{1}{2}|A(\mathbf{q})|\mathbf{q}_{xx}\,\Delta x} \qquad (4.11)$$

It takes into account the viscosity effects of the *spatial discretisation*. The matrix on the right hand side is called viscosity matrix of the Roe scheme

$$V_{\text{Roe}} = |A(\mathbf{q})|$$

in analogy to the viscosity terms in the Navier-Stokes equation.

Characteristic form

We want to analyse the artificial viscosity on the individual characteristic waves. For this reason we derive the characteristic form of the modified equation.

Let $\mathbf{r}_1, \ldots, \mathbf{r}_4$ be the right eigenvectors of the Jacobian $A = \mathrm{d}\mathbf{f}/\mathrm{d}\mathbf{q}$ at the state $\mathbf{q}(x,t)$ with the eigenvalues

$$\lambda_1 = u - a, \quad \lambda_2 = u, \quad \lambda_3 = u, \quad \lambda_4 = u + a.$$

We apply the transformation matrices R and R^{-1} defined in (4.2) and (4.2) to the modified Equation (4.11)

$$R^{-1}\mathbf{q}_t + (R^{-1}AR)R^{-1}\mathbf{q}_x = \frac{1}{2}(R^{-1}|A|R)R^{-1}\mathbf{q}_{xx}\Delta x$$

and, using the relations $|A| = R|\Lambda|R^{-1}$ and $\mathrm{d}\mathbf{w} = R^{-1}\mathrm{d}\mathbf{q}$, we obtain the modified equation in characteristic variables

$$\mathbf{w}_t + \Lambda\mathbf{w}_x = \frac{1}{2}|\Lambda|\mathbf{w}_{xx}\Delta x. \qquad (4.12)$$

Evaluating the ratio of the physical convection to the artificial dissipation we obtain the numerical Reynolds number for all characteristic waves:

$$\boxed{\mathrm{Re}_{\mathrm{Roe}} = \frac{1}{\Delta x}}$$

Since the ratio is independent of the Mach number we expect no excessive artificial viscosity in the limit $M \to 0$ originating from the spatial discretisation, i.e. from Roe's approximate Riemann solver.

4.2.2. Explicit time discretisation

A simple first-order temporal discretisation is the (forward) Euler method. We obtain its modified expression by replacing the time derivative by the discrete expression and then using Taylor expansions:

$$\mathbf{q}_t \rightsquigarrow \frac{\mathbf{q}^{n+1} - \mathbf{q}^n}{\Delta t} = \mathbf{q}_t + \frac{1}{2}\mathbf{q}_{tt}\Delta t + \mathcal{O}(\Delta t^2) \,.$$

In order to express $\mathbf{q}_{tt}$ by $\mathbf{q}_{xx}$ we use the modified equation except for first-order terms:

$$\mathbf{q}_t \doteq -A\mathbf{q}_x \,. \tag{4.13}$$

Note that first-order terms will turn to second-order terms in the final modified equation and can therfore be neglected. The spatial derivative of (4.13) is

$$\mathbf{q}_{tx} \doteq -A_x \mathbf{q}_x - A\mathbf{q}_{xx} \,,$$

which can be used in the temporal derivative of (4.13) to obtain

$$\begin{aligned}
\mathbf{q}_{tt} &\doteq -A_t \mathbf{q}_x - A\mathbf{q}_{xt} \\
&\doteq -A_t \mathbf{q}_x - A(-A_x \mathbf{q}_x - A\mathbf{q}_{xx}) \\
&\doteq -A_t \mathbf{q}_x + AA_x \mathbf{q}_x + A^2 \mathbf{q}_{xx} \\
&\doteq -A_\mathbf{q} \mathbf{q}_t \mathbf{q}_x + AA_\mathbf{q} \mathbf{q}_x^2 + A^2 \mathbf{q}_{xx} \\
&\doteq A_\mathbf{q} A \mathbf{q}_x^2 + AA_\mathbf{q} \mathbf{q}_x^2 + A^2 \mathbf{q}_{xx} \,,
\end{aligned} \tag{4.14}$$

where we have expressed the spatial and temporal derivatives of the Jacobian A by the product of the third-order tensor $A_\mathbf{q}$ and the respective derivatives of the state vector $\mathbf{q}$:

$$A_x = A_\mathbf{q} \mathbf{q}_x \,, \qquad A_t = A_\mathbf{q} \mathbf{q}_t \,.$$

The modified equation for the discrete Roe scheme is therefore

$$\mathbf{q}_t + \big(A + \underbrace{\frac{A_\mathbf{q} A \mathbf{q}_x + AA_\mathbf{q} \mathbf{q}_x}{2}\Delta t}_{\Delta A}\big)\mathbf{q}_x = \frac{1}{2}|A|\mathbf{q}_{xx}\Delta x - \frac{1}{2}A^2 \mathbf{q}_{xx}\Delta t \,. \tag{4.15}$$

The temporal discretisation not only introduces a viscosity but also changes the Jacobian, with two consequences: the wave speeds and eigenvectors change so that a complete diagonalisation of the modified equation is no longer possible. Nevertheless, we use the eigenvectors of the original Jacobian A to introduce the characteristic variables $\mathrm{d}\mathbf{w} = R^{-1}\,\mathrm{d}\mathbf{q}$ introduced for the semi-discrete system.

Characteristic form

If we do so we obtain a coupled system of equations

$$\mathbf{w}_t + \Lambda \mathbf{w}_x + R^{-1}\Delta AR\mathbf{w}_x = \frac{1}{2}|\Lambda|\mathbf{w}_{xx}\Delta x - \frac{1}{2}\Lambda^2 \mathbf{v}_{xx}\Delta t$$

that can also be written as

$$\mathbf{w}_t + (\Lambda + \Delta)\mathbf{w}_x = \frac{1}{2}|\Lambda|(I-C)\mathbf{w}_{xx}\Delta x ,$$

where $C = \mathrm{diag}(c_1, \ldots, c_4)$ is the diagonal matrix of Courant numbers $c_i = \lambda_i \Delta t / \Delta x$. The diagonal elements of the perturbing matrix $\Delta = R^{-1}\Delta AR$ change the wave speeds of the original system by $\mathcal{O}(\Delta t)$, while the off-diagonal elements are responsible for a coupling, which is $\mathcal{O}(\Delta t)$. We neglect the contribution to the original Jacobian for small time steps. The ratio of convective transport to artificial viscosity is then

$$\boxed{\mathrm{Re}_{\mathrm{Roe,ex}} = \frac{1}{(1-c_i)\Delta x}}$$

for all waves. For the acoustic waves the Courant numbers $c_\pm$ are set to ≈ 0.9 for explicit schemes, while entropy and shear wave have Courant numbers of order $\mathcal{O}_S(\mathrm{M})$. Thus the dissipation of the latter is stronger than for the acoustic waves but the order of the numerical Reynolds number is $\mathcal{O}_S(1/\Delta x)$ as $\mathrm{M} \to 0$ and therfore independent of the Mach number for all characteristic wave equations. This has an interesting effect on the numerical behaviour: since the Roe scheme is consistent, there is a grid on which the numerical results for moderately small Mach number flow are good approximations of inviscid flow. On this grid we can lower the Mach number without risking excessive artificial viscosity that might deteriorate the accuracy. In numerical experiments we have slowed down the flow on a given grid to Mach numbers as low as $\mathrm{M} = 10^{-14}$ without loss of accuracy. Note, that in this test series we have used a quad precision compilation of the code to avoid the *cancellation problem*.

4.2.3. Numerical results

General remark on the choice of test cases

As one-dimensional test cases could serve simple Riemann problems consisting of a single jump in the density, creating only simple entropy waves, or a single jump in the transverse velocity, creating only simple shear waves. The dissipation of contact discontinuities is, more or less, independent of the Mach number for the Roe scheme. We do not present any results of one-dimensional test cases here.

More interesting are test cases in 2D, which allow a direct comparison with results presented in the literature [60, 36, 6]. As two-dimensional test cases we choose the steady flow around a cylinder and around a NACA0012 aerofoil. Note that in Part II of this thesis it is shown that the accuracy problem is avoided on triangular finite volume cells for the Roe scheme. We therefore assume: if, nevertheless, the accuracy problem occurs for a first-order upwind scheme, it can be entirely ascribed to the properties of the used flux solver. Therefore we think it is legitimate to use *two-dimensional* test cases (with triangular grids) to verify the numerical behaviour analysed in a *one-dimensional* setting.

General remark on initial and boundary conditions

The settings valid for almost all test studies in this thesis are given below. Any deviations thereof are given in the specific context of the numerical experiment.

The *initial conditions* are uniform

$$\rho_0 = 1.0,$$
$$\mathbf{u}_0 = (u_0, v_0)^T,$$
$$p_0 = 1.0,$$

where the absolute value of the initial velocity

$$\|\mathbf{u}_0\| = \sqrt{u_0^2 + v_0^2}$$

is set to meet the prescribed initial Mach number

$$\mathrm{M}_0 = \frac{\|\mathbf{u}_0\|}{a_0} = \frac{\|\mathbf{u}_0\|}{\sqrt{\gamma p_0/\rho_0}} = \frac{\|\mathbf{u}_0\|}{\sqrt{\gamma}},$$

where γ is the adiabatic index set to 1.4 throughout our calculations. The angle of attack (relevant for NACA0012) is defined as

$$\alpha_0 = \tan\frac{v_0}{u_0}.$$

The exact solution at infinity is assumed uniform (free-stream values)

$$\rho_\infty = 1.0,$$
$$\mathbf{u}_\infty = (u_\infty, v_\infty)^T,$$
$$p_\infty = 1.0.$$

In the *far-field boundary conditions* this solution is assumed to be a good approximation of the solution at the outer boundary. To this end the computational domain is embedded by a layer of ghost cells in which we impose the solution at infinity:

$$\rho_{\text{ghost}} = 1.0,$$
$$\mathbf{u}_{\text{ghost}} = (u_\infty, v_\infty)^T,$$
$$p_{\text{ghost}} = 1.0.$$

This is a good approximation whenever the external boundary is far from obstacles inside the domain. The upwind-character of the scheme automatically takes into account, whether the flow is entering (inflow) or leaving the domain (outflow) at an interface on the boundary. Note that in all our calculations initial conditions and the conditions at infinity are set equal with the consequences, that

$$\text{M}_\infty = \text{M}_0 \quad \text{and} \quad \mathbf{u}_\infty = \mathbf{u}_0.$$

For this reason we refer to far-field, inflow or initial Mach number with the same symbol M_0. Note, in conjunction with order relations we use the symbol M defined in *analytical* contexts as $\text{M} = u_{\text{ref}}/a_{\text{ref}}$ and in *numerical* contexts as $\text{M} = u_0/a_0 = u_\infty/a_\infty$.

Entropy production and transport

In *inviscid flow* there is no entropy production except at shocks. In low Mach number flows, i.e. without shocks, entropy is a conserved quantity that is transported with the flow (isentropic flow). Under certain initial and boundary conditions entropy is constant throughout the entire flow domain for all times (homentropic flow). In *viscous flow* however, entropy is generated due to energy and momentum dissipation, for example in acoustic waves or within regions with shear layers such as boundary layers.

First-order upwind schemes for the inviscid Euler equations introduce artificial viscosity for stability reasons and produce entropy as a side effect. The amount of entropy produced and the way it is transported is a direct indicator for the quality of the numerical scheme. We investigated the long time behaviour of the Roe scheme with regard to entropy production and transport. For the numerical experiment we have used the following specific settings:

Grid: An unstructured grid around a circular disc is used with 9822 triangular grid cells, see Figure 4.4 for a detail. Similar results are obtained for structured grids with triangular cells. All calculations are done on the primary grid.

Discretisation: first-order in space and time with Roe flux, Courant number $c = 0.5$ (2D) and global time steps.

In Figure 4.1 the contour lines of the entropy fluctuation

$$\tilde{s} = c_V \ln \frac{p}{p_0} - c_p \ln \frac{\rho}{\rho_0}$$

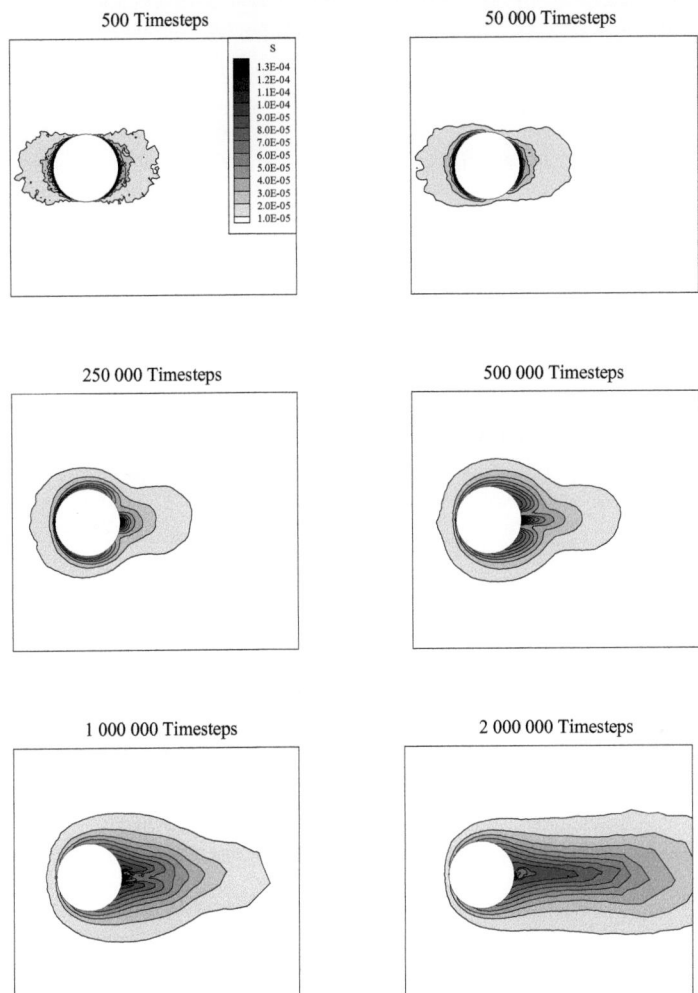

Figure 4.1.: Entropy production due to artificial viscosity and its transport by the explicit first-order Roe scheme for the flow around a cylinder at $M_0 = 10^{-3}$. Shown are the contour lines of the entropy fluctuation $\tilde{s}$.

are shown for different numbers of time steps – all with the same contour level settings for better comparison. After 500 time steps the entropy distribution is symmetric. Its origin lies in the energy dissipation of the strong pressure wave produced in the initialisation process: the homogeneous distribution of initial flow velocity leads to a strong divergence of the flow field at the stagnation point in the beginning of the calculation. In the following pictures this entropy is transported in downwind direction and additional entropy is continuously produced. For the location of the entropy production we can make out two regions: the stagnation points on the upwind and the downwind side of the cylinder. The reason thereof is the shear flow in the vicinity of the stagnation points, where the flow velocity vanishes. Calculations up to 10 million time steps show a balance between entropy production and convective transport leading to a constant entropy in the flow domain accumulated in a tail-shaped region behind the cylinder. The flow in this region can start to swirl like a von Kármán vortex street. In anticipation of the HLL scheme we want to emphasise here: the Roe scheme transports entropy by *convection* and not by *numerical dissipation*.

Vorticity production and transport

Inviscid flow conserves vorticity along particle paths. If there is no vorticity introduced by the initial or boundary conditions, then no vorticity should be created. Again, *artificial viscosity* of numerical schemes causes the deviation from ideal flow, so that the vorticity produced is an indicator for the accuracy of the scheme. For the long term study of this effect we used the same settings as before. In Figure 4.2 the contour lines of the vorticity

$$\omega = \text{curl } \mathbf{u}$$

are shown with the same colour map for all time steps. We see that the vorticity is produced near the surface of the cylinder and transported with the flow.

Comparison with an analytical solution

For the flow around a cylinder we compare the incompressible potential flow solution as given in Section 2.3.3 with the numerical approximation by the Roe scheme. For the simulation we used the settings as before except for variations:

Grid: unstructured grid with 140 000 triangular cells.

Mach Number: $M_\infty = M_0 = 10^{-6}$

Figure 4.3 shows lines of constant pressure fluctuation

$$\tilde{p} = \frac{p - p_\infty}{p_\infty}$$

for the incompressible potential flow around a cylinder (left) at an inflow velocity of $u_\infty = \sqrt{\gamma}\, 10^{-6}$ and the results obtained with the first-order Roe scheme (right) at an inflow Mach number of $M_0 = 10^{-6}$. We mention that the background pressure was

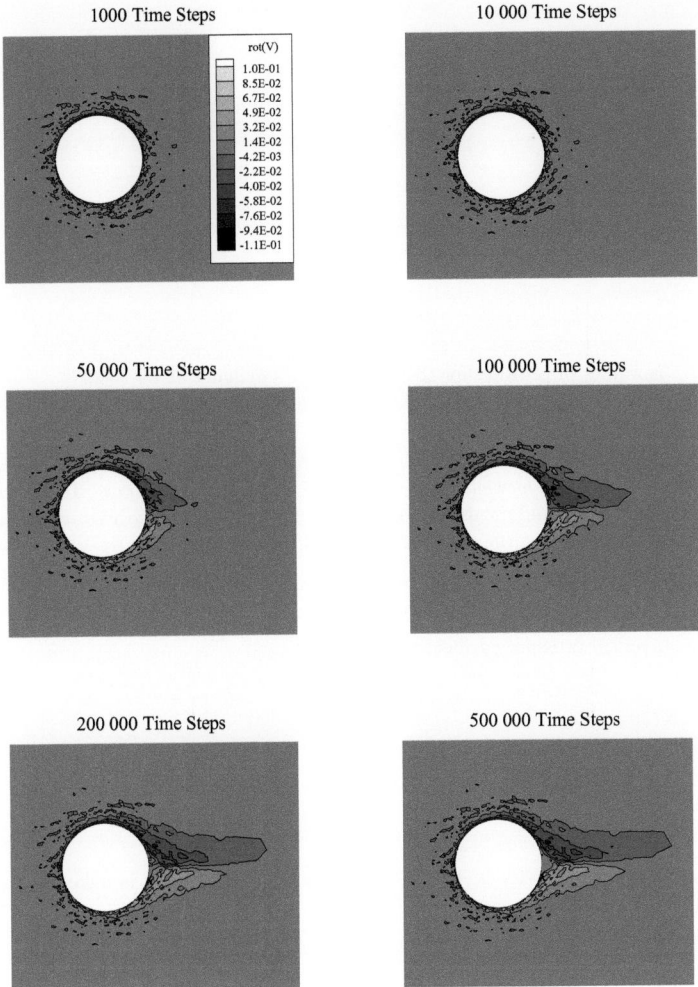

Figure 4.2.: Vorticity production and transport for the flow around a cylinder at $M_0 = 10^{-3}$ with the explicit first-order Roe scheme. Shown are the contour lines of the vorticity $\omega = \operatorname{curl} \mathbf{u}$.

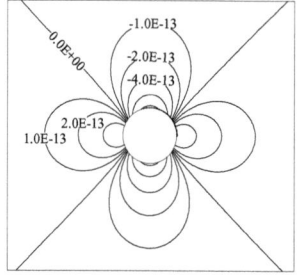 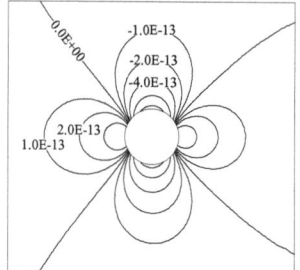

Figure 4.3.: Isovalues of pressure for the flow around a cylinder at $M_0 = 10^{-6}$.
Left: incompressible potential flow. Right: first-order Roe scheme.

scaled to unity $p_\infty = 1$ in all our calculations. Both pressure distributions have the same structure of isolines, as well as the same magnitude of the variations. To see the agreement between the analytical solution and the numerical result, in Figure 4.4 the pressure coefficient at the surface for the analytical solution

$$c_p = \frac{p - p_\infty}{\frac{1}{2}\rho_\infty u_\infty^2} = 2\cos 2\phi - 1$$

and the numerical pressure coefficient:

$$c_{p,\text{num}} = \frac{p - p_\infty}{\frac{1}{2}\gamma M_\infty^2},$$

are shown for comparison in a diagram. Note that the numerical pressure variations are scaled with the kinetic energy at the inflow

$$\frac{1}{2}\rho_\infty u_\infty^2 = \frac{1}{2}\gamma M_\infty^2,$$

where the background density is $\rho_\infty = 1$ and the inflow velocity u_∞ is linked to the Mach number for ideal gases by

$$M_\infty = \frac{u_\infty}{a_\infty} = \frac{u_\infty}{\sqrt{\gamma p_\infty / \rho_\infty}},$$

with the background pressure scaled to $p_\infty = 1$.

Flow around a NACA0012 aerofoil

As a standard test case for compressible solvers we chose the flow around a NACA0012 aerofoil with the following settings specific to this test case:

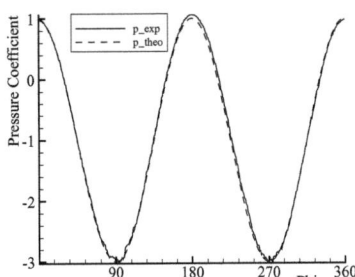

Figure 4.4.: Left: detail of cylinder grid with 9822 cells.
Right: pressure coefficient at the cylinder surface of the incompressible potential flow solution (dashed line) and for the explicit first-order Roe scheme (solid line) at $u_\infty/\sqrt{\gamma} = M_0 = 10^{-6}$.

Grid: An unstructured NACA0012 grid with 2068 cells is used, see Figure 4.5 for a detail of the grid.

Discretisation: explicit first-order Roe scheme, CFL number $c = 0.5$, forward Euler temporal discretisation.

Far-field Boundary: free-stream Mach Number $M_\infty = 10^{-6}$, angle of attack $\alpha = 0$.

In Figure 4.5 a detail of the grid (left) and the isovalues of the pressure (right) are shown. The shape and position of the pressure isolines are in good agreement with the literature [60].

Magnitude of pressure variations

For some upwind schemes the observed pressure fluctuations show ill-shaped isolines as well as a wrong order of magnitude $\mathcal{O}_S(M)$ instead of $\mathcal{O}_S(M^2)$. We tested the explicit first-order Roe scheme with the flow around a NACA0012 aerofoil and the flow around a cylinder with various refinements. The maximal pressure fluctuation

$$p_{\text{fluc}} = \frac{p_{\max} - p_{\min}}{p_{\max}}$$

was independent of the grid. The results for the NACA0012 with 2068 cells is shown in Figure 4.6. The pressure-Mach number curve coincides with the function $f(M) = M^2$, which is physically correct. The results can be verified with the data given in Table 4.1.

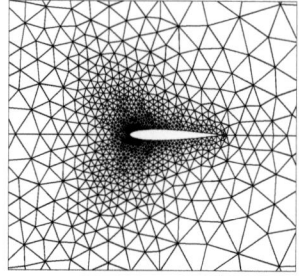 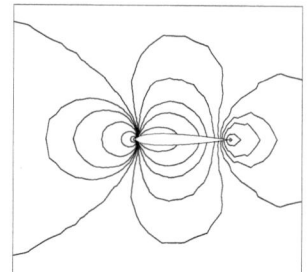

Figure 4.5.: Left: detail of the NACA0012 aerofoil grid with 2068 cells
Right: isolines of pressure for the flow around a NACA0012 aerofoil at $M_0 = 10^{-6}$ with the first-order Roe scheme.

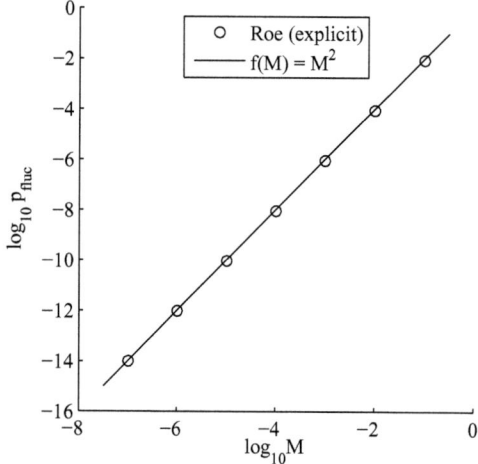

Figure 4.6.: Maximal pressure fluctuation $p_{\text{fluc}} = (p_{\max} - p_{\min})/p_{\max}$ against inflow Mach number for the explicit first-order Roe scheme.

M_0	p_{fluc}	$(p_{\text{fluc}} - M_0^2)/p_{\text{fluc}}$
1.0094063430234E-01	0.87436230925623E-02	0.17
1.0094947513878E-02	0.88517806831499E-04	0.15
1.0137451388955E-03	0.89201481342395E-06	0.15
1.0192892135121E-04	0.91382100188020E-08	0.14
1.0206937811147E-05	0.94109942055216E-10	0.11
1.0208755753697E-06	0.97011287891678E-12	0.07
1.0208952606303E-07	0.99920072216264E-14	0.04

Table 4.1.: Relative deviation between p_{fluc} and M_0^2 for the flow around the NACA0012 aerofoil with the explicit first-order Roe scheme.

4.2.4. Implicit time discretisation

The contribution of the spatial discretisation to the artificial viscosity is known from the previous section. Now the focus is on the viscous effects of the implicit temporal discretisation.

We choose the simple backward Euler time discretisation and check, whether the additional artificial viscosity – especially for large times steps – leads to Mach number dependent numerical Reynolds numbers. The implicit Roe scheme with backward Euler can be written as

$$\frac{\mathbf{q}_i^{n+1} - \mathbf{q}_i^n}{\Delta t} = -\frac{1}{\Delta x}(\mathbf{f}_{i+1/2}^{n+1} - \mathbf{f}_{i-1/2}^{n+1}) , \qquad (4.16)$$

where the Roe fluxes are evaluated at the new time level t_{n+1}. We expand the smooth solution $\mathbf{q}(x,t)$ about the time level t_{n+1} in a Taylor series and obtain for the time derivative of the LHS of (4.16):

$$\frac{\mathbf{q}_i^{n+1} - \mathbf{q}_i^n}{\Delta t} = \mathbf{q}_t(x_i, t_{n+1}) - \frac{1}{2}\mathbf{q}_{tt}(x_i, t_{n+1})\Delta t + \mathcal{O}(\Delta t^2) . \qquad (4.17)$$

Substituting (4.17) in (4.16) and using the modified equation for the spatial discretisation (4.11) we obtain

$$\mathbf{q}_t + A(\mathbf{q})\mathbf{q}_x = \frac{1}{2}|A(\mathbf{q})|\mathbf{q}_{xx}\Delta x + \frac{1}{2}\mathbf{q}_{tt}\Delta t . \qquad (4.18)$$

We make use of (4.14) to eliminate the second-order time derivative in (4.18)

$$\mathbf{q}_t + \big(A - \underbrace{\frac{A_{\mathbf{q}}A\mathbf{q}_x + AA_{\mathbf{q}}\mathbf{q}_x}{2}\Delta t}_{\Delta A}\big)\mathbf{q}_x = \frac{1}{2}|A|\mathbf{q}_{xx}\Delta x + \frac{1}{2}A^2\mathbf{q}_{xx}\Delta t .$$

After transforming to the characteristic variables of the semi-discrete Roe scheme, we obtain

$$\mathbf{w}_t + (\Lambda - \Delta)\mathbf{w}_x = \frac{1}{2}|\Lambda|(I + C)\mathbf{w}_{xx}\Delta x .$$

As for the explicit implementation there remains a part $\Delta = R^{-1}\Delta A R$ that is not diagonalisable. This term is responsible for a change of wave speeds and for a weak coupling of the equations of order $\mathcal{O}(\Delta t)$. Neglecting these effects, approximate numerical Reynolds numbers can be derived for all characteristic waves:

$$\boxed{\text{Re}_{\text{Roe,im}} \approx \frac{1}{(1+c_i)\Delta x}}$$

where $c_i = \lambda_i \Delta t / \Delta x$ are the Courant numbers. At first glance everything seems unproblematic, since the numerical Reynolds numbers are $\mathcal{O}_S(1/\Delta x)$. If we choose time steps on the flow time scale, this remains true for the shear and entropy wave, but the Courant numbers of the acoustic waves become $\mathcal{O}_S(1/M)$. As a consequence, the numerical Reynolds numbers for the acoustic waves change their order relation:

$$\boxed{\text{Re}_{\text{Roe,im}}^{\pm} \approx \frac{1}{(1+c_{\pm})\Delta x} \sim \frac{M}{\Delta x} \quad \text{as} \quad M \to 0} \qquad (4.19)$$

In such an implementation the acoustic waves are more damped then transported as $M \to 0$.

Numerical results

Numerical results for the behaviour of the implicit Roe scheme in the low Mach number regime are presented by Viozat and Guillard in [60, 20]. As a matter of fact, large time steps are not feasible in reality, since Newton's method does not converge for large time steps too large and (4.19) is just of theoretical interest.

4.3. Summary

We will now summarise the results of this chapter by extending the concept of *asymptotic consistency* of upwind methods for the linear advection equation to Godunov-type schemes. This definition builds on the characteristic form of the modified equation and is therefore restricted to the one-dimensional model case.

Definition 4.3.1. *A first-order upwind scheme for the Euler equations is called **asymptotically consistent**, if for all characteristic variables the corresponding modified equation has a numerical Reynolds number which satisfies*

$$\text{Re}_{\text{num}} = \mathcal{O}_S\left(\frac{1}{\Delta x}\right) \quad as \quad M \to 0 \,.$$

If for a characteristic variable the modified equation satisfies

$$\text{Re}_{\text{num}} = \mathcal{O}_S\left(\frac{M}{\Delta x}\right) \quad as \quad M \to 0 \,,$$

*the scheme is called **asymptotically inconsistent with respect to this characteristic wave**.*

Theorem 4.3.2. *The following variants of the Roe scheme are **asymptotically consistent**:*

- *the explicit Roe scheme (cf. Section 4.2)*
- *the implicit Roe scheme with small time steps, i.e. Courant numbers of order $\mathcal{O}_S(1)$ for the acoustic waves and $\mathcal{O}_S(M)$ for the entropy and shear wave (cf. Section 4.2.4)*

*The following variant of the Roe scheme is **asymptotically inconsistent** with respect to the acoustic waves:*

- *the implicit Roe scheme with large time steps, i.e. Courant numbers of order $\mathcal{O}_S(1/M)$ for the acoustic waves and $\mathcal{O}_S(1)$ for the entropy and shear wave (cf. Section 4.2.4)*

Proof. The proof for the diverse variants of the Roe scheme can be found in the corresponding sections. □

We conclude the analysis of the first-order Roe scheme with a suggestion: *If an upwind scheme does not show the accuracy problem (on triangular finite volume cells), it resolves all characteristic waves, i.e. the artificial viscosity on all characteristic waves remains small compared to the advection of the characteristic variable for small Mach numbers.*

5. The HLL scheme

In this section we analyse the asymptotic behaviour of numerical schemes that are based on the two-wave Riemann solver by Harten, Lax and van Leer (HLL) [21, 51]. A modification of HLL that resolves contact or shear waves called HLLC, cf. [51], will be shown to avoid the excessive dissipation problem for HLL. For this purpose we want to derive the modified equation of the HLL Riemann solver in characteristic variables to see the viscosity effects on the different characteristic waves of the two-dimensional Euler equations.

Another type of HLL which resolves all characteristic waves in the Riemann problem (HLLEM) was proposed by Einfeldt in [14]. It is not explicitly analysed here but the numerical results are identical to the ones obtained with the Roe scheme.

5.1. The two-wave HLL scheme

As with the Roe scheme, we investigate the numerical viscosity of the spatial and temporal discretisations separately, starting with the semi-discrete form.

5.1.1. Spatial discretisation

The first-order HLL scheme in semi-discrete form is given by

$$\frac{\mathrm{d}}{\mathrm{d}t}\mathbf{q}_i = -\frac{1}{\Delta x}(\mathbf{f}^{\mathrm{HLL}}_{i+1/2} - \mathbf{f}^{\mathrm{HLL}}_{i-1/2}),$$

with the numerical flux function as given in [24]

$$\mathbf{f}_{\mathrm{HLL}}(\mathbf{q}_\mathrm{R}, \mathbf{q}_\mathrm{L}) = \frac{\mathbf{f}(\mathbf{q}_\mathrm{R}) + \mathbf{f}(\mathbf{q}_\mathrm{L})}{2} - \frac{1}{2}\frac{S_\mathrm{R} + S_\mathrm{L}}{S_\mathrm{R} - S_\mathrm{L}}(\mathbf{f}(\mathbf{q}_\mathrm{R}) - \mathbf{f}(\mathbf{q}_\mathrm{L})) + \frac{S_\mathrm{R} S_\mathrm{L}}{S_\mathrm{R} - S_\mathrm{L}}(\mathbf{q}_\mathrm{R} - \mathbf{q}_\mathrm{L}), \quad (5.1)$$

under the assumption that

$$S_\mathrm{L} \leq 0 \leq S_\mathrm{R}, \quad S_\mathrm{L} \neq S_\mathrm{R},$$

which is justified in the low Mach number regime. Here S_L and S_R are wave speeds that are faster than the left- and right-running signal speeds in the underlying Riemann problem and have to be estimated adequately. If there is a Roe-Matrix $\hat{A} = \hat{A}(\mathbf{q}_\mathrm{R}, \mathbf{q}_\mathrm{L})$ for this Riemann problem we can write

$$\mathbf{f}(\mathbf{q}_\mathrm{R}) - \mathbf{f}(\mathbf{q}_\mathrm{L}) = \hat{A}(\mathbf{q}_\mathrm{R} - \mathbf{q}_\mathrm{L}).$$

Substituting this into Equation (5.1) we obtain

$$\mathbf{f}_{\text{HLL}}(\mathbf{q}_\text{R}, \mathbf{q}_\text{L}) = \frac{\mathbf{f}(\mathbf{q}_\text{R}) + \mathbf{f}(\mathbf{q}_\text{L})}{2} - \frac{1}{2}\frac{S_\text{R} + S_\text{L}}{S_\text{R} - S_\text{L}} \hat{A} (\mathbf{q}_\text{R} - \mathbf{q}_\text{L}) + \frac{S_\text{R} S_\text{L}}{S_\text{R} - S_\text{L}} (\mathbf{q}_\text{R} - \mathbf{q}_\text{L})$$

for the numerical flux of the HLL scheme. We assume $\mathbf{q}(x,t)$ to be a smooth solution to the original PDE. Following the approach of Section 4.2 we obtain for the modified equation of the HLL scheme

$$\mathbf{q}_t + A\mathbf{q}_x = \frac{1}{2} V_{\text{HLL}} \mathbf{q}_{xx} \Delta x \, ,$$

where

$$V_{\text{HLL}} = \frac{S_\text{R} + S_\text{L}}{S_\text{R} - S_\text{L}} |A(\mathbf{q})| - 2\frac{S_\text{R} S_\text{L}}{S_\text{R} - S_\text{L}} I + \mathcal{O}(\Delta x)$$

is the viscosity matrix of the HLL scheme with the Jacobian $A(\mathbf{q})$ and the identity matrix I as presented in [24].

Characteristic form

The viscosity matrix V_{HLL} has the same eigenvectors $\mathbf{r}_1, \ldots, \mathbf{r}_4$ as the Jacobian $A = \mathrm{d}\mathbf{f}/\mathrm{d}\mathbf{q}$, so that we can transform the modified equation using $R = [\mathbf{r}_1, \ldots, \mathbf{r}_4]$:

$$R^{-1}\mathbf{q}_t + (R^{-1}AR)R^{-1}\mathbf{q}_x = \frac{1}{2}(R^{-1}V_{\text{HLL}}R)R^{-1}\mathbf{q}_{xx}\Delta x \, ,$$

into a (locally) decoupled system with the characteristic variables $\mathrm{d}\mathbf{w} = R^{-1}\mathrm{d}\mathbf{q}$:

$$\mathbf{w}_t + \Lambda \mathbf{w}_x = \frac{1}{2}\Lambda_{\text{HLL}} \mathbf{w}_{xx} \Delta x \, , \tag{5.2}$$

where $\Lambda = \mathrm{diag}(u-a, u, u, u+a)$ is the matrix of eigenvalues and

$$\Lambda_{\text{HLL}} = R^{-1}\Big(\underbrace{\frac{S_\text{R}+S_\text{L}}{S_\text{R}-S_\text{L}}}_{\alpha}|A| - \underbrace{2\frac{S_\text{R} S_\text{L}}{S_\text{R}-S_\text{L}}I}_{\delta}\Big)R \tag{5.3}$$

is the characteristic form of the viscosity matrix V_{HLL}. It can be simplified using the relation $|A| = R|\Lambda|R^{-1}$ and the identity $R^{-1}IR = I$ to obtain

$$\Lambda_{\text{HLL}} = \alpha|\Lambda| - \delta I$$
$$= \mathrm{diag}(\alpha|u-a| - \delta, \alpha|u| - \delta, \alpha|u| - \delta, \alpha|u+a| - \delta)$$
$$=: \mathrm{diag}(\mu_1, \ldots, \mu_4) \, ,$$

where the μ_i can be interpreted as *numerical viscosity coefficients* for the individual characteristic waves. The numerical Reynolds numbers, as the ratios of convective transport to numerical dissipation of the characteristic variables, are given by

$$\boxed{\mathrm{Re}_{\text{HLL}} = \frac{|\lambda_i|}{\mu_i \Delta x}} \tag{5.4}$$

In many implementations of the HLL scheme the wave speeds satisfy the asymptotic equalities
$$\begin{aligned} S_R &\sim u + a \\ S_L &\sim u - a \end{aligned} \quad \text{as} \quad M \to 0.$$

With this assumption we find for α
$$\alpha = \frac{S_R + S_L}{S_R - S_L} \sim \frac{2u}{2a} = M \tag{5.5}$$

and for δ
$$\delta = 2\frac{S_R S_L}{S_R - S_L} \sim \frac{(u-a)(u+a)}{a},$$

which can be written as
$$\delta \sim (M - 1)(u + a) \tag{5.6a}$$

or equivalently as
$$\delta \sim (M + 1)(u - a). \tag{5.6b}$$

The coefficient of artificial viscosity μ_1 for the *left-running acoustic wave* satisfies
$$\begin{aligned} \mu_1 &= \alpha|u - a| - \delta \\ &\sim M|u - a| - (u - a)(M + 1) \\ &\sim (2M + 1)(a - u). \end{aligned}$$

Inserting this and $|\lambda_1| = a - u$ into (5.4) gives for the corresponding numerical Reynolds number
$$\text{Re}_{\text{HLL}}^{(1)} \sim \frac{1}{2M+1}\frac{1}{\Delta x} \sim \frac{1}{\Delta x} \quad \text{as} \quad M \to 0.$$

For the right-running acoustic wave we obtain
$$\begin{aligned} \mu_4 &= \alpha|u + a| - \delta \\ &\sim M|u + a| - |M - 1|(u + a) \\ &= u + a \end{aligned}$$

and with $|\lambda_4| = u + a$
$$\text{Re}_{\text{HLL}}^{(4)} \sim \frac{1}{\Delta x} \quad \text{as} \quad M \to 0.$$

For contact and shear wave with $|\lambda_{2,3}| = |u|$ the coefficients of artificial viscosity are
$$\begin{aligned} \mu_{2,3} &= \alpha|u| - \delta \\ &\sim M|u| - (M - 1)(u + a), \end{aligned}$$

so that we obtain for the numerical Reynolds number

$$\mathrm{Re}_{\mathrm{HLL}}^{(2,3)} \sim \frac{|u|}{\mathrm{M}|u| - (\mathrm{M}-1)(u+a)} \frac{1}{\Delta x}$$
$$\sim \frac{1}{\mathrm{M} + (\mathrm{M}-1)(\pm 1 + \frac{1}{\mathrm{M}})} \frac{1}{\Delta x}$$
$$\sim \frac{\mathrm{M}}{\Delta x}$$

as $\mathrm{M} \to 0$. The results are summarised in the following box

$$\boxed{\mathrm{Re}_{\mathrm{HLL}}^{(1,4)} \sim \frac{1}{\Delta x} \quad \text{and} \quad \mathrm{Re}_{\mathrm{HLL}}^{(2,3)} \sim \frac{\mathrm{M}}{\Delta x} \quad \text{as} \quad \mathrm{M} \to 0}$$

This states that the Riemann solver damps the acoustic waves independently of the Mach number for $\mathrm{M} \to 0$, whereas for the entropy and shear wave the situation is different: compared to the physical transport, the artificial dissipation grows like $1/\mathrm{M}$ for $\mathrm{M} \to 0$.

5.1.2. Explicit time discretisation

The time derivative is discretised with forward Euler and introduces a further dissipation term into the modified equation:

$$\mathbf{q}_t \rightsquigarrow \frac{\mathbf{q}^{n+1} - \mathbf{q}^n}{\Delta t} = \mathbf{q}_t + \frac{1}{2}\mathbf{q}_{tt}\Delta t + \mathcal{O}(\Delta x^2) \, .$$

Following the approach used for the Roe scheme in Chapter 4 we eliminate the second-order temporal derivative and obtain for the modified equation for the fully discrete HLL scheme

$$\mathbf{q}_t + (A + \Delta A)\mathbf{q}_x = \frac{1}{2}V_{\mathrm{HLL}}\mathbf{q}_{xx}\Delta x - \frac{1}{2}A^2\mathbf{q}_{xx}\Delta t \, ,$$

where ΔA is defined in (4.15). In the characteristic variables of the original Jacobian A this can be written as

$$\mathbf{w}_t + (\Lambda + \Delta)\mathbf{w}_x = \frac{1}{2}\Lambda_{\mathrm{HLL}}\mathbf{w}_{xx}\Delta x - \frac{1}{2}\Lambda^2 \mathbf{w}_{xx}\Delta t$$
$$= \frac{1}{2}\mathbf{w}_{xx}\Delta x(\Lambda_{\mathrm{HLL}} - \Lambda C) \, ,$$

where $C = \mathrm{diag}(c_1, \ldots, c_4)$ is a diagonal matrix with the Courant numbers on the diagonal and $\Delta = R^{-1}\Delta A R$. Neglecting Δ, which is $\mathcal{O}(\Delta t)$, we obtain an approximation for the numerical Reynolds numbers

$$\boxed{\mathrm{Re}_{\mathrm{HLL,ex}} \approx \frac{|\lambda_i|}{(\mu_i - \lambda_i c_i)\Delta x}}$$

The order relation for the numerical Reynolds numbers of the acoustic waves can be derived using (5.5) for α and (5.6) for δ, as well as the fact $c_{1,4} = \mathcal{O}_S(1)$, to obtain

$$\text{Re}^{(1,4)} \sim \frac{|u \pm a|}{(\alpha|u \pm a| - \delta - (u \pm a)c_{1,4})\Delta x} = \mathcal{O}_S\left(\frac{1}{\Delta x}\right).$$

For the entropy and shear wave the order relation is

$$\text{Re}^{(2,3)} \sim \frac{|u|}{(\alpha|u| - \delta - uc_{2,3})\Delta x} = \mathcal{O}_S\left(\frac{M}{\Delta x}\right),$$

where we have used the fact that $c_{2,3} = \mathcal{O}_S(M)$. Obviously the temporal discretisation does not change the order relation of the numerical Reynolds numbers obtained for the semi-discrete form of the HLL scheme.

5.1.3. Numerical results

We use the same test cases and settings as for the Roe scheme described in Section 4.2.3.

Entropy production and transport

We first check the HLL scheme with respect to the production and transport of entropy. In Figure 5.1 the entropy is shown for different numbers of time steps – all with the same contour level settings for better comparison. We can make out the two stagnation points as the centres of entropy production. The spatial distribution of the entropy is isotropic, i.e. the direction of advection is not represented. In comparison with the explicit Roe scheme, HLL scheme has a large entropy production and is characterised by a *dissipative transport*. An analogy in physics is given by the Navier-Stokes equations for very low physical Reynolds numbers and is called *Stokes or creeping flow*.

Vorticity production and transport

In Figure 5.2 it can be seen that the production of vorticity takes place near the upper and lower side of the cylinder. The vorticity transport is decoupled from the advection direction and is rather isotropic, so that the underlying mechanism is *dissipation*.

Comparison with the analytical solution

We use the flow around a cylinder to compare the pressure fields of the numerical results and the incompressible potential flow solution. In Figure 5.3 the isovalues of the pressure fluctuation

$$\tilde{p} = \frac{p - p_\infty}{p_\infty}$$

are shown for both cases. The results obtained with the HLL schemes deviate from the analytical solution in several aspects:

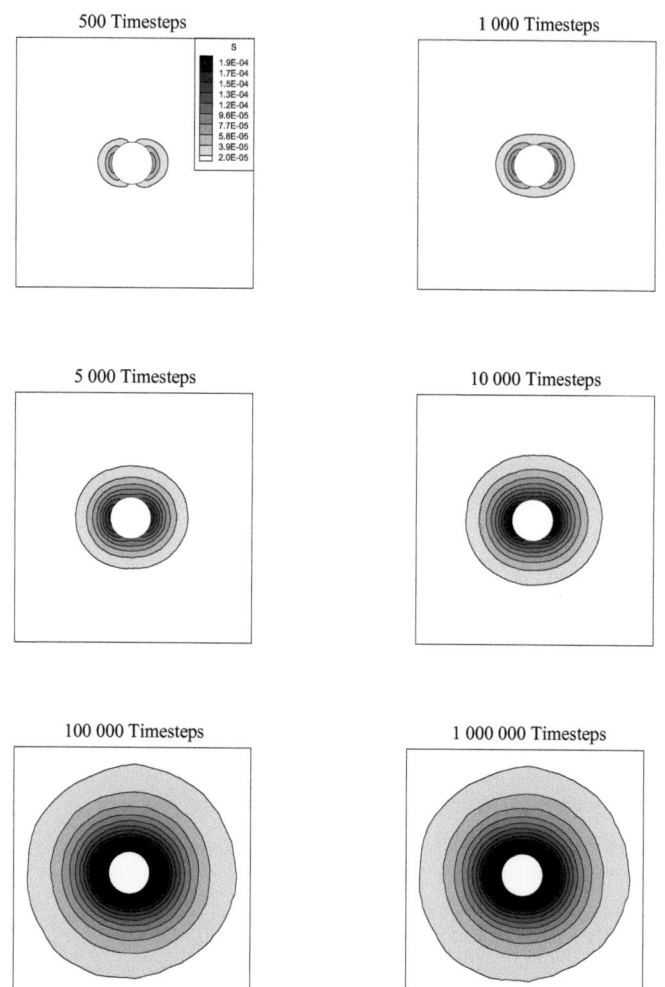

Figure 5.1.: Entropy production due to artificial viscosity by the first-order HLL scheme for the flow around a cylinder at $M_0 = 10^{-3}$. Shown are the contour lines of the entropy fluctuation $\tilde{s}$, cf. Equation (2.6), with a fixed grey scale.

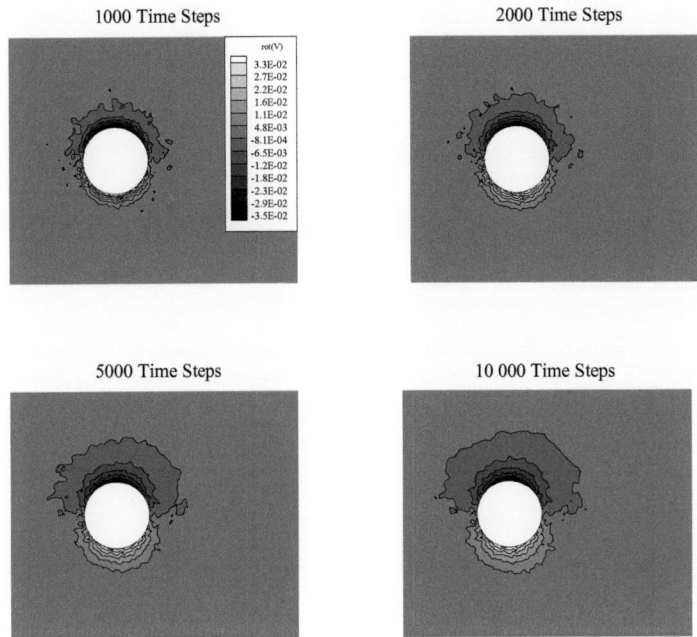

Figure 5.2.: Vorticity production by the first-order HLL scheme for the flow around a cylinder at $M_0 = 10^{-3}$. Shown are the contour lines of the vorticity $\omega = \text{curl } \mathbf{u}$ with a fixed grey scale.

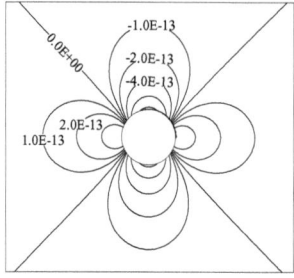 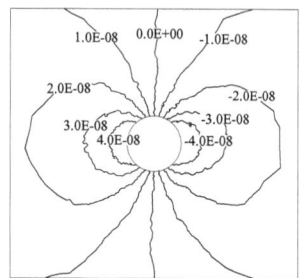

Figure 5.3.: Isovalues of pressure for the flow around a cylinder at $M_0 = 10^{-6}$. Left: incompressible potential flow. Right: explicit first-order HLL.

- The shape of the pressure field is completely different. Instead of two high pressure regions at the front and back of the cylinder and two regions of low pressure at the upper and lower side, there are only two regions: high pressure in the upwind and low pressure in the downwind direction.

- The magnitude of the pressure in the stagnation point is 10^5 times larger than in the analytical solution.

- The lines of constant pressure are no longer smooth curves.

We observed similar results for inflow Mach numbers between $M_0 = 10^{-1}$ and $M_0 = 10^{-12}$.

Wrong magnitude of pressure fluctuations

Diagram 5.4 shows the maximal pressure fluctuation

$$p_{\text{fluc}} = \frac{p_{\max} - p_{\min}}{p_{\max}}$$

obtained with the HLL scheme for different Mach numbers. Obviously the pressure difference is of the wrong order $\mathcal{O}_S(M)$ instead of $\mathcal{O}_S(M^2)$.

Analogy to creeping flow

The numerical results for the entropy production and transport by the HLL scheme suggests a parallel to a limit case of the Navier-Stokes equation: in *creeping* or *Stokes*

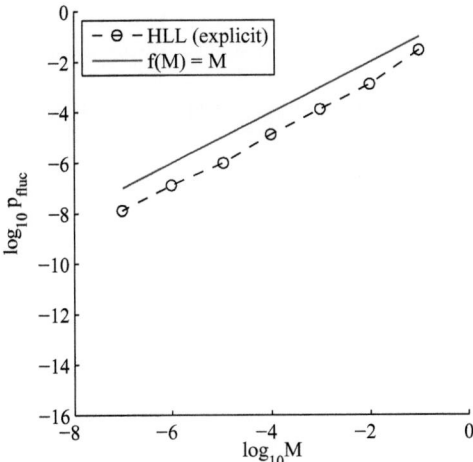

Figure 5.4.: Maximal pressure fluctuation $p_{\text{fluc}} = (p_{\max} - p_{\min})/p_{\max}$ against the inflow Mach number M for the flow around a cylinder with the first-order HLL scheme.

flow the transport of momentum by physical dissipation exceeds the convective transport by orders of magnitude and transforms the Navier-Stokes system into a diffusion dominated equation, cf. Section 2.2.2 for details. Creeping flow is characterised by a diminishing (physical) Reynolds numbers

$$\text{Re} \to 0,$$

which is imitated by numerical schemes with a numerical Reynolds number for the shear wave of the form

$$\text{Re}_{\text{num}}^{\text{ShearWave}} = \mathcal{O}_S\left(\frac{M}{\Delta x}\right) \quad \text{as} \quad M \to 0$$

in the low Mach number limit. We have to point out here that only the numerical viscosity on the shear wave is comparable to the physical viscosity in the momentum equation of viscous flow.

Since there are analytical approximations to creeping flow around a sphere or a cylinder we can validate this analogy. But there is a slight difference between creeping flow and numerical flow simulation with vanishing numerical Reynolds numbers: the wall boundary condition in the Navier-Stokes equation are no-slip wall boundary conditions $u_\parallel = u_\perp = 0$, whereas in numerical schemes for inviscid flow simulations we impose the less restrictive kinematic wall boundary condition $u_\perp = 0$. If we want to compare the numerical results we have to modify the analytical approximation of the creeping flow equations, given in text books on theoretical hydromechanics such as [29], by introducing the (unphysical) kinematic wall boundary condition. The solution is then no longer unique but the order relation can be expected to be the same. One possible solution for the pressure near the surface of the cylinder is given by:

$$\boxed{p_{\text{Stokes}} = p_\infty - \rho u_\infty A_0 \frac{\cos\theta}{r}} \qquad (5.7)$$

In Figure 5.5 the isolines of the pressure fluctuation $\tilde{p}_{\text{Stokes}} = p_{\text{Stokes}} - p_\infty$ for creeping flow (left) can be compared with the variations of the calculated pressure

$$\tilde{p} = \frac{p - p_\infty}{p_\infty},$$

obtained with the HLL scheme (right). The physical viscosity for the creeping flow solution was set to $\eta = 0.25$ in order to fit the results obtained with HLL on the given grid. The shape of the isolines are in good agreement: the distributions are both symmetric to the x-axis and show a region of high pressure in the upwind and a low pressure region in the downwind direction. Slight deviation are further away from the cylinder, where the approximation (5.7) is no longer valid. The good agreement between creeping flow and HLL solution can also be seen in Figure 5.6, where we compare the pressure fluctuation $\tilde{p}$ on the surface of the cylinder. Note, the angle $\phi = 180°$ corresponds to the stagnation point in the upwind direction.

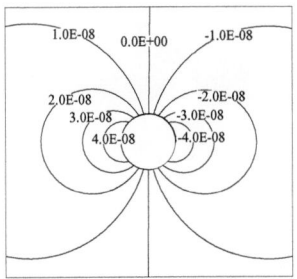 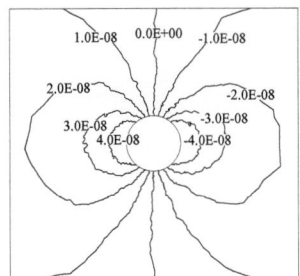

Figure 5.5.: Isolines of pressure for the flow around a cylinder at $M_0 = 10^{-3}$. Left: creeping flow. Right: first-order HLL.

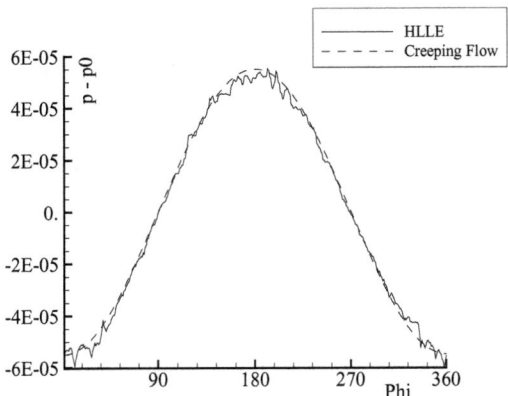

Figure 5.6.: Surface pressure for the flow around a cylinder at $M_0 = 10^{-3}$. Dashed line: creeping flow. Solid line: first-order HLL.

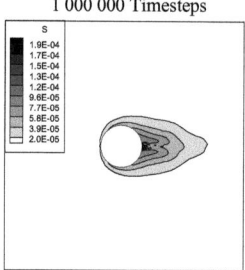
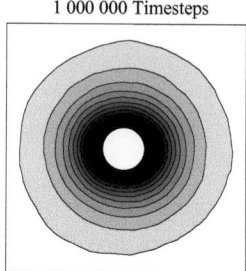

Figure 5.7.: Steady entropy distribution for the flow around a cylinder at $M_0 = 10^{-3}$. Left: first-order Roe scheme. Right: first-order HLL.

5.1.4. Roe and HLL – a comparison

In this section we summarise the major results by showing the different behaviour of the explicit upwind schemes by Roe and by Harten, Lax and van Leer in the low Mach number regime. The analysis of the modified equation is summarised in the following diagram symbolising the numerical behaviour for $M \to 0$:

$$\mathrm{Re}_{\mathrm{HLL}}^{(\|)} = \mathcal{O}_S\Big(\frac{M}{\Delta x}\Big) \quad \longleftrightarrow \quad \mathrm{Re}_{\mathrm{Roe}}^{(\|)} = \mathcal{O}_S\Big(\frac{1}{\Delta x}\Big)$$

$$\Downarrow \qquad\qquad\qquad \Downarrow$$

$$\text{asymptotically inconsistent} \qquad \text{asymptotically consistent}$$

$$\Downarrow \qquad\qquad\qquad \Downarrow$$

$$\text{creeping flow} \qquad\qquad \text{potential flow}$$

Although we are approximating the solution of the inviscid Euler equations, the artificial viscosity in the HLL scheme is increasing with lower Mach numbers. As a consequence the numerical entropy production of the HLL scheme is much higher than for the Roe scheme and, more importantly, the transport is dominated by *dissipation* as opposed to *convection* for the Roe scheme. This effect is shown in Figure 5.7, where the same grey scale is used for both plots. In Figure 5.8 the isolines of pressure for the flow around a cylinder are shown. The structure of the pressure distribution is entirely different: Roe approximates a potential flow solution with $\tilde{p} = \mathcal{O}_S(M^2)$ and the HLL scheme approximates creeping flow with $\tilde{p} = \mathcal{O}_S(M)$. While the flow around a cylinder is good for comparing numerical results with analytical solutions,

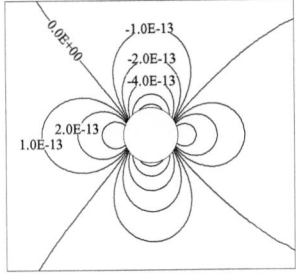 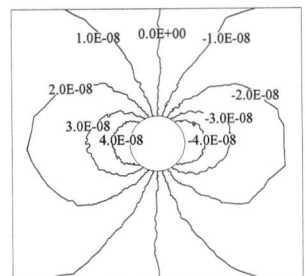

Figure 5.8.: Isolines of pressure for the flow around cylinder at $M_0 = 10^{-6}$. Left: first-order Roe scheme. Right: first-order HLL.

the NACA012 aerofoil is a wide spread test case for compressible solvers. The isolines of pressure for the flow around this aerofoil obtained with the explicit Roe (left) and HLL scheme are given in Figure 5.9. As before, the magnitude of pressure is correct for Roe but not for HLL. The shape of the isolines of pressure obtained with the two upwind schemes deviates, but the inaccuracy of the distribution obtained with the HLL scheme is less evident. It becomes even more difficult to say for an angle of attack different from zero as can be seen at the bottom of Figure 5.9, where the angle of attack was set to $\alpha = 5°$ and the inflow Mach number was set to $M_0 = 10^{-2}$. The subtle difference to the correct solution is the absence of a high pressure region in the wake of the aerofoil characterising potential flow..

5.1.5. Implicit time discretisation

For completeness we analyse the HLL scheme with an implicit time discretisation. In analogy to the Roe scheme a temporal discretisation with the backward Euler method leads to an additional term on the RHS of the modified equation:

$$\mathbf{q}_t + A(\mathbf{q})\mathbf{q}_x = \frac{1}{2}V_{\text{HLL}}\mathbf{q}_{xx}\Delta x + \frac{1}{2}\mathbf{q}_{tt}\Delta t \ . \tag{5.8}$$

Expressing the time derivative on the RHS of (5.8) by spatial derivatives leads to

$$\mathbf{q}_t + (A - \Delta A)\mathbf{q}_x = \frac{1}{2}V_{\text{HLL}}\mathbf{q}_{xx}\Delta x + \frac{1}{2}A^2\mathbf{q}_{xx}\Delta x \ ,$$

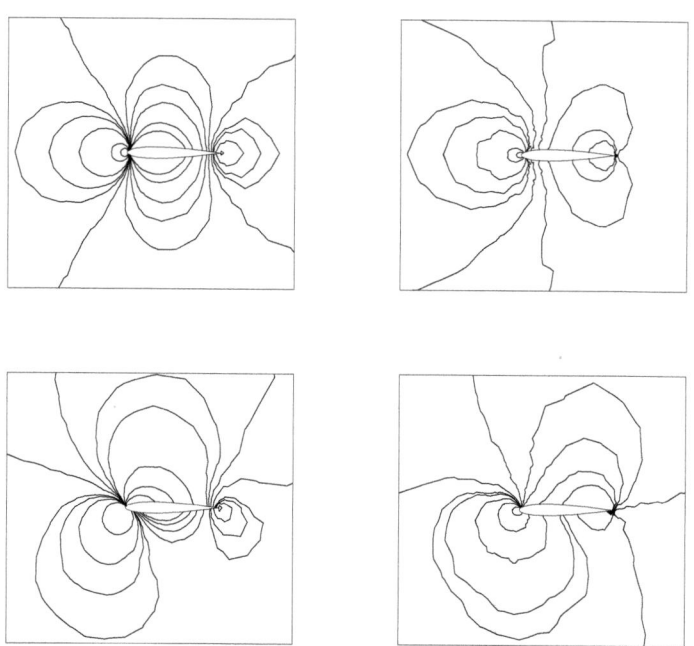

Figure 5.9.: Isolines of pressure for the flow around a NACA0012 aerofoil at $M_0 = 10^{-2}$ with $\alpha = 0°$ (top) and $\alpha = 5°$ (bottom) angle of attack.
Left column: first-order Roe scheme. Right column: first-order HLL.

where the perturbation matrix ΔA is given by (4.15). This equation can be written in characteristic variables $d\mathbf{w} = R^{-1} d\mathbf{q}$ of the original Jacobian A

$$\mathbf{w}_t + (\Lambda - \Delta)\mathbf{q}_x = \frac{1}{2}\Lambda_{\text{HLL}} \mathbf{w}_{xx} \Delta x + \frac{1}{2}\Lambda^2 \mathbf{w}_{xx} \Delta t$$

$$= \frac{1}{2}(\Lambda_{\text{HLL}} + \Lambda C)\mathbf{w}_{xx} \Delta x ,$$

where $C = \text{diag}(c_1, \ldots, c_4)$ is the diagonal matrix of Courant numbers. If we neglect the $\mathcal{O}(\Delta t)$-perturbation of the physical flux Δ, we obtain for the numerical Reynolds numbers

$$\boxed{\text{Re}_{\text{HLL,im}} \approx \frac{|\lambda_i|}{(\mu_i + \lambda_i c_i)\Delta x}}$$

Using $\alpha = \mathcal{O}_S(M)$, $\delta = \mathcal{O}_S(a)$ and

$$c_{1,4} = \mathcal{O}_S\left(\frac{1}{M}\right),$$

valid for time steps on the flow time scale, we can derive the following order relation for the acoustic waves:

$$\text{Re}_{\text{HLL,im}}^{(1,4)} \approx \frac{|u \pm a|}{(\alpha|u \pm a| - \delta + (u \pm a)c_{1,4})\Delta x} = \mathcal{O}_S\left(\frac{M}{\Delta x}\right) \quad \text{as} \quad M \to 0 ,$$

and using $c_{2,3} = \mathcal{O}_S(1)$, for the entropy and shear waves

$$\text{Re}_{\text{HLL,im}}^{(2,3)} \approx \frac{|u|}{(\alpha|u| - \delta + uc_{2,3})\Delta x} = \mathcal{O}_S\left(\frac{M}{\Delta x}\right) \quad \text{as} \quad M \to 0 .$$

The temporal discretisation with its large Courant numbers of order $\mathcal{O}_S(1/M)$ are responsible for the excessive dissipation of the acoustic waves, while the excessive dissipation of the shear and entropy waves is due to the term δ, which is related to the spatial discretisation.

5.2. HLLC

The HLLC scheme, presented in [3] for example, explicitly resolves contact waves unlike the original two-wave HLL Riemann solver. Unfortunately, the dissipation matrix V_{HLLC} of HLLC and the Jacobian A have different eigenspaces, i.e. decoupled advection-diffusion equations for shear and entropy waves do not exist for a general Riemann problem. To overcome this difficulty, we use a special Riemann problem, where the initial data has only a jump in the transverse velocity, i.e. an explicitly given shear wave. This enables us to analyse the dissipative behaviour of the numerical scheme for this isolated characteristic wave in a simple way. To set up the modified equation, two parallel Riemann problems of this kind are necessary to derive the two fluxes at the cell interfaces.

5.2.1. Transport of a shear wave

HLLC is given for the one-dimensional model problem by

$$\frac{d}{dt}\mathbf{q}_i + \frac{1}{\Delta x}(\mathbf{f}^{\text{HLLC}}_{i+1/2} - \mathbf{f}^{\text{HLLC}}_{i-1/2}) = 0 \,,$$

with the HLLC flux as given in [51]

$$\mathbf{f}^{\text{HLLC}}_{i+1/2} = \frac{1 + \sigma_{i+1/2}}{2}\{\mathbf{f}_i + S^{\text{L}}_{i+1/2}(\mathbf{q}^{*\text{L}}_{i+1/2} - \mathbf{q}_i)\}$$
$$+ \frac{1 - \sigma_{i+1/2}}{2}\{\mathbf{f}_{i+1} + S^{\text{R}}_{i+1/2}(\mathbf{q}^{*\text{R}}_{i+1/2} - \mathbf{q}_{i+1})\} \,,$$

where $\sigma_{i+1/2} = \text{sign}(S^*_{i+1/2})$ is the sign of the estimated speed of the contact wave and

$$\mathbf{q}^{*\text{K}} = \rho_{\text{K}}\frac{S_{\text{K}} - u_{\text{K}}}{S_{\text{K}} - S_*}\begin{bmatrix} 1 \\ S_* \\ v_{\text{K}} \\ \frac{E_{\text{K}}}{\rho_{\text{K}}} + (S_* - u_{\text{K}})\left[S_* + \frac{p_{\text{K}}}{\rho_{\text{K}}(S_{\text{K}}-u_{\text{K}})}\right] \end{bmatrix} ,$$

with K = L and K = R, are the two HLLC states. In the simplified situation of a pure shear flow all physical quantities apart from the transverse velocity v are constant:

$$u = \text{const}, \quad p = \text{const}, \quad \rho = \text{const}, \quad a = \text{const},$$

leading to some helpful simplifications for the HLLC flux:

$$S_* = u\,, \quad S_{\text{L}} = u - a\,, \quad S_{\text{R}} = u + a\,, \quad \sigma = \text{sign}(S_*) = \text{sign}(u)\,.$$

The flux difference for the momentum ρv can be written as

$$f^{\rho v}_{i+1/2} - f^{\rho v}_{i-1/2} = \tfrac{1}{2}(1+\sigma)\rho u v_i + \tfrac{1}{2}(1-\sigma)\rho u v_{i+1} - \tfrac{1}{2}(1+\sigma)\rho u v_{i-1} - \tfrac{1}{2}(1-\sigma)\rho u v_i$$

$$\begin{aligned}
&= \tfrac{1}{2}\rho u v_i \quad + \tfrac{1}{2}\sigma\rho u v_i \\
&\;+ \tfrac{1}{2}\rho u v_{i+1} - \tfrac{1}{2}\sigma\rho u v_{i+1} \\
&\;- \tfrac{1}{2}\rho u v_{i-1} - \tfrac{1}{2}\sigma\rho u v_{i-1} \\
&\;- \tfrac{1}{2}\rho u v_i \quad + \tfrac{1}{2}\sigma\rho u v_i,
\end{aligned}$$

$$= \tfrac{1}{2}\rho u \Delta_{i+1,i-1} v - \tfrac{1}{2}\sigma\rho u \Delta^2_i v \,,$$

where we have used the discrete operators

$$\Delta_{i,j} v := v_i - v_j \,, \tag{5.9}$$

$$\Delta^2_i v := v_{i+1} - 2v_i + v_{i-1} \,, \tag{5.10}$$

and we omitted the indices for constant quantities. Using the modified equation approach for the transverse velocity v, we obtain

$$\frac{\partial}{\partial t}\rho v + u\frac{\partial}{\partial x}\rho v = \frac{1}{2}\sigma u\frac{\partial^2}{\partial x^2}\rho v \Delta x + \mathcal{O}(\Delta x^2) \ .$$

The numerical Reynolds number for the shear wave transport is therefore

$$\boxed{\mathrm{Re}_{\mathrm{HLLC}} = \frac{|u|}{\sigma u \Delta x} = \mathcal{O}_{\mathrm{S}}\!\left(\frac{1}{\Delta x}\right) \quad \text{as} \quad \mathrm{M} \to 0}$$

which is of the correct order of magnitude, i. e. independent of the Mach number.

5.2.2. Numerical results

Comparison with analytical solution

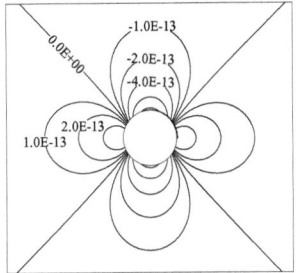 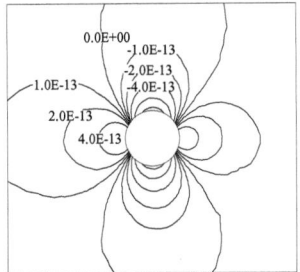

Figure 5.10.: Isolines of pressure for the flow around a cylinder at $M_0 = 10^{-6}$.
Left: incompressible potential flow. Right: first-order HLLC.

In Figure 5.10 we present the pressure fluctuation $\tilde{p}_{\mathrm{HLLC}} = (p_{\mathrm{HLLC}} - p_\infty)/p_\infty$ obtained with the HLLC scheme with the same settings as used with the Roe scheme in Section 5.1.3. Further calculations with other Mach numbers confirm the relation

$$\tilde{p}_{\mathrm{HLLC}} = \mathcal{O}_{\mathrm{S}}(\mathrm{M}^2) \ .$$

The result suggests that the resolution of the contact wave is the key to avoiding the accuracy problem of Godunov-type schemes in the low Mach number regime.

5.3. Summary

Since we could not show analytically that

$$\text{Re}_{\text{num}} = \mathcal{O}_{\text{S}}\left(\frac{1}{\Delta x}\right)$$

for the acoustic and the entropy wave we have to make a minor restriction for the HLLC scheme in the following theorem.

Theorem 5.3.1. *The following variants of the HLL scheme are* **asymptotically consistent** *with respect to the shear wave:*

- *the explicit HLLC scheme (cf. Section 5.2),*

- *the explicit HLLEM scheme as a special case of HLLC with S_L, S_R chosen according to [24].*

The following variants of the HLL scheme are **asymptotically inconsistent** *with respect to the shear wave:*

- *the explicit HLL scheme (cf. Section 5.1),*

- *the implicit HLL scheme with small time steps, i.e. Courant numbers of order $\mathcal{O}(1)$ for the acoustic waves and $\mathcal{O}(\text{M})$ for the entropy and shear waves (cf. Section 5.1.5),*

- *the Rusanov scheme as a special case of the explicit HLL with $S_\text{L} = S_\text{R}$ and*

- *the Lax-Friedrichs scheme as a special case of the explicit HLL with $S_\text{L} = S_\text{R} = \Delta x / \Delta t$.*

The implicit HLL with large time steps, i.e. Courant numbers of order $\mathcal{O}_\text{S}(1/\text{M})$ for the acoustic waves and $\mathcal{O}_\text{S}(1)$ for the entropy and shear wave, is **asymptotically inconsistent** *with respect to all characteristic waves.*

Proof. The proof for the diverse variants of the HLL scheme can be found in the corresponding sections. □

6. Flux vector splitting methods

In this chapter the *accuracy problem* is analysed for various flux vector splitting methods in the context of the one-dimensional model case described in Section 4.1.

As for HLLC, the flux vector splitting methods have dissipation matrices that cannot be diagonalised with the Jacobian, i. e. a decoupled advection-diffusion equation for the shear and entropy wave does not exist for a general Riemann problem. We therefore explicitly set up a shear wave by using a special Riemann problem where the initial data has only a jump in the transverse velocity. This allows to analyse the dissipative behaviour of the schemes on this characteristic wave. Explicit analysis and numerical results confirming the theory are presented for the flux vector splittings of van Leer, Steger-Warming and for the advection upstream method (AUSM) by Liou and Steffen. The representation of the schemes follows the book [51] by E. Toro.

6.1. The van Leer splitting

In the one-dimensional model situation, cf. Equation (4.1), the van Leer splitting can be written as

$$\mathbf{q}_t + \frac{1}{\Delta x}(\mathbf{f}^{\text{vL}}_{i+1/2} - \mathbf{f}^{\text{vL}}_{i+1/2}) = 0 \;,$$

where the flux vector splitting is defined by

$$\mathbf{f}^{\text{vL}}_{i+1/2} = \mathbf{f}^{+}_{i} + \mathbf{f}^{-}_{i+1} \;, \qquad \mathbf{f}^{\text{vL}}_{i-1/2} = \mathbf{f}^{+}_{i-1} + \mathbf{f}^{-}_{i} \;,$$

with the flux of the vertical momentum ρv given by

$$f^{+}_{\rho v} = \frac{1}{4}\rho a (1+\text{M})^2 v \;, \qquad f^{-}_{\rho v} = -\frac{1}{4}\rho a (1-\text{M})^2 v \;.$$

$\text{M} = u/a$ is the local Mach number with respect to the normal velocity u in the Riemann problem. Since in a pure shear wave all quantities but the transverse velocity v are constant, the momentum fluxes at the interfaces can be written as

$$f^{\rho v}_{i+1/2} = \frac{1}{4}\rho a \left\{ (1+\text{M})^2 v_i - (1-\text{M})^2 v_{i+1} \right\} \;,$$

$$f^{\rho v}_{i-1/2} = \frac{1}{4}\rho a \left\{ (1+\text{M})^2 v_{i-1} - (1-\text{M})^2 v_i \right\} \;,$$

so that we obtain for the flux difference:

$$\begin{aligned} f^{\rho v}_{i+1/2} - f^{\rho v}_{i-1/2} = \tfrac{1}{4}\rho a\{ & v_i & + & 2\mathrm{M}v_i & + & \mathrm{M}^2 v_i \\ & -v_{i+1} & + & 2\mathrm{M}v_{i+1} & - & \mathrm{M}^2 v_{i+1} \\ & -v_{i-1} & - & 2\mathrm{M}v_{i-1} & - & \mathrm{M}^2 v_{i-1} \\ & +v_i & - & 2\mathrm{M}v_i & + & \mathrm{M}^2 v_i \} \end{aligned}$$

$$= \tfrac{1}{4}\rho a\{ 2\mathrm{M}\Delta_{i+1,i-1}v - (1+\mathrm{M}^2)\Delta_i^2 v \} \,,$$

where we have used the discrete difference operators Δ and Δ^2 as defined in Equation (5.9) and (5.9). Let v be a smooth solution to the original PDE, then we obtain for the transport of the shear wave the following modified equation

$$\frac{\partial}{\partial t}\rho v + u \frac{\partial}{\partial x}\rho v = \frac{1+\mathrm{M}}{4}\frac{\partial^2}{\partial x^2}\rho v \Delta x + \mathcal{O}(\Delta x^2) \,.$$

The numerical Reynolds number for the transport of a pure shear wave is therefore

$$\boxed{\mathrm{Re}_{\mathrm{vanLeer}} = \frac{2u}{a(1+\mathrm{M}^2)\Delta x} = \mathcal{O}_{\mathrm{S}}\!\left(\frac{\mathrm{M}}{\Delta x}\right) \quad \text{as} \quad \mathrm{M} \to 0}$$

Obviously, the numerically induced dissipation far outweighs the convective transport of the shear wave for small Mach numbers. We conclude:

The first-order van Leer splitting shows the accuracy problem in the low Mach number regime due to excessive dissipation of shear waves in the local Riemann problems.

6.2. The Steger-Warming splitting

The split fluxes for the transport of the momentum ρv as given in [51], can be written for the Steger-Warming splitting as

$$f^{\pm}_{\rho v} = \frac{\rho}{2\gamma}(v\lambda_1^{\pm} + 2(\gamma-1)v\lambda_2^{\pm} + v\lambda_4^{\pm}) \,,$$

where, in low Mach number flow, the characteristic speeds satisfy

$$\lambda_1 = u - a < 0 \,,$$
$$\lambda_{2,3} = u \lessgtr 0 \,,$$
$$\lambda_4 = u + a > 0 \,.$$

Using the constancy of all quantities in the shear flow except for the transverse velocity v, we obtain for the fluxes at the interfaces

$$f^{\rho v}_{i+1/2} = \frac{\rho}{2\gamma}\{[2(\gamma-1)u^+ + (u+a)]v_i + [2(\gamma-1)u^- + (u-a)]v_{i+1}\} \,,$$

$$f^{\rho v}_{i-1/2} = \frac{\rho}{2\gamma}\{[2(\gamma-1)u^+ + (u+a)]v_{i-1} + [2(\gamma-1)u^- + (u-a)]v_i\} \,,$$

where
$$u^+ = \frac{u+|u|}{2}, \quad u^- = \frac{u-|u|}{2}.$$

For the flux difference we obtain

$$\begin{aligned}
f^{\rho v}_{i+1/2} - f^{\rho v}_{i-1/2} &= \tfrac{\rho}{2\gamma}\{ 2(\gamma-1)u^+v_i && + uv_i && + av_i \\
&\quad + 2(\gamma-1)u^-v_{i+1} && + uv_{i+1} && - av_{i+1} \\
&\quad - 2(\gamma-1)u^+v_{i-1} && - uv_{i-1} && - av_{i-1} \\
&\quad - 2(\gamma-1)u^-v_i && - uv_i && + av_i \} \\
&= \tfrac{\rho}{2\gamma}\{ (\gamma-1)uv_i && + (\gamma-1)|u|v_i && + uv_i && + av_i \\
&\quad + (\gamma-1)uv_{i+1} && - (\gamma-1)|u|v_{i+1} && + uv_{i+1} && - av_{i+1} \\
&\quad - (\gamma-1)uv_{i-1} && - (\gamma-1)|u|v_{i-1} && - uv_{i-1} && - av_{i-1} \\
&\quad - (\gamma-1)uv_i && + (\gamma-1)|u|v_i && - uv_i && + av_i \} \\
&= \tfrac{\rho}{2\gamma}\{ \gamma uv_i && + (\gamma-1)|u|v_i && + av_i \\
&\quad + \gamma uv_{i+1} && - (\gamma-1)|u|v_{i+1} && - av_{i+1} \\
&\quad - \gamma uv_{i-1} && - (\gamma-1)|u|v_{i-1} && - av_{i-1} \\
&\quad - \gamma uv_i && + (\gamma-1)|u|v_i && + av_i \} .
\end{aligned}$$

Using the discrete difference operators defined in (5.9) and (5.9), we obtain for the flux difference:

$$\frac{f^{\rho v}_{i+1/2} - f^{\rho v}_{i-1/2}}{\Delta x} = \rho u \frac{\Delta_{i+1,i-1} v}{2\Delta x} - \frac{1}{2}\frac{\rho[(\gamma-1)|u|+a]}{\gamma}\frac{\Delta_i^2 v}{\Delta x}.$$

This can be used to derive the modified equation for the transport of the momentum ρv:

$$\frac{\partial}{\partial t}\rho v + u\frac{\partial}{\partial x}\rho v = \frac{1}{2}\frac{(\gamma-1)|u|+a}{\gamma}\frac{\partial^2}{\partial x^2}\rho v \Delta x + \mathcal{O}(\Delta x^2) .$$

For the numerical Reynolds number of the shear wave we obtain

$$\boxed{\mathrm{Re}_{\text{Steger-Warming}} = \frac{\gamma u}{(\gamma-1)|u|+a}\frac{1}{\Delta x} = \mathcal{O}_{\mathrm{S}}\!\left(\frac{\mathrm{M}}{\Delta x}\right) \quad \text{as} \quad \mathrm{M}\to 0}$$

This explains the *accuracy problem* of the first-order Steger-Warming splitting in the low Mach number regime.

6.3. The Liou-Steffen scheme (AUSM)

In the Liou and Steffen flux vector splitting, cf. [51], the x-split two-dimensional flux is separated into a convective and a pressure component:

$$\mathbf{f} = \begin{pmatrix} \rho u \\ \rho u^2 \\ \rho u v \\ \rho u h \end{pmatrix} + \begin{pmatrix} 0 \\ p \\ 0 \\ 0 \end{pmatrix} = \mathrm{M} \begin{pmatrix} \rho a \\ \rho a u \\ \rho a v \\ \rho a h \end{pmatrix} + \begin{pmatrix} 0 \\ p \\ 0 \\ 0 \end{pmatrix} = \mathbf{f}^{(c)} + \mathbf{f}^{(p)} \,,$$

where $\mathrm{M} = u/a$ is the local Mach number according to the velocity normal to the Riemann problem, and h is the specific total enthalpy. The *convective flux component* is given by

$$\mathbf{f}^{(c)}_{i+1/2} = \mathrm{M}_{i+1/2} \begin{cases} \mathbf{f}^c_i & \text{if } \mathrm{M}_{i+1/2} \geq 0 \,, \\ \mathbf{f}^c_{i+1} & \text{if } \mathrm{M}_{i+1/2} \leq 0 \,. \end{cases}$$

The cell-interface Mach number $\mathrm{M}_{i+1/2}$ is given by the splitting

$$\mathrm{M}_{i+1/2} = \mathrm{M}^+_i + \mathrm{M}^-_{i+1} \,,$$

with

$$\mathrm{M}^\pm = \begin{cases} \pm \tfrac{1}{4}(\mathrm{M} \pm 1)^2 & \text{if } |\mathrm{M}| \leq 1 \,, \\ \tfrac{1}{2}(\mathrm{M} \pm |\mathrm{M}|) & \text{if } |\mathrm{M}| > 1 \,. \end{cases}$$

For the splitting of the pressure

$$p_{i+1/2} = p^+_i + p^-_{i+1}$$

we used

$$p^\pm = \begin{cases} \tfrac{1}{2} p (1 \pm \mathrm{M}) & \text{if } |\mathrm{M}| \leq 1 \,, \\ \tfrac{1}{2} p \frac{\mathrm{M} \pm |\mathrm{M}|}{\mathrm{M}} & \text{if } > 1 \,. \end{cases}$$

We assume $|\mathrm{M}| < 1$, valid for low Mach number flow regions, so that we obtain for the cell-interface values of Mach number and pressure:

$$\mathrm{M}_{i+1/2} = \tfrac{1}{4}(\mathrm{M}_i + 1)^2 - \tfrac{1}{4}(\mathrm{M}_{i+1} - 1)^2 \,,$$

$$p_{i+1/2} = \tfrac{1}{2} p_i (1 + \mathrm{M}_i) + \tfrac{1}{2} p_{i+1} (1 - \mathrm{M}_{i+1}) \,.$$

In the special situation of a pure shear wave we have

$$\rho = \text{const} \,, \quad u = \text{const} \,, \quad a = \text{const} \,, \quad \mathrm{M} = \text{const} \,.$$

To start with, we assume $\mathrm{M} < 0$ and obtain for the flux of the momentum ρv transverse to the Riemann problem:

$$f^{\rho v}_{i+1/2} = \underbrace{\{\tfrac{1}{4}(\mathrm{M}+1)^2 - \tfrac{1}{4}(\mathrm{M}-1)^2\}}_{\mathrm{M}} \rho a v_{i+1} = \rho u v_{i+1} \,,$$

$$f^{\rho v}_{i-1/2} = \underbrace{\{\tfrac{1}{4}(\mathrm{M}+1)^2 - \tfrac{1}{4}(\mathrm{M}-1)^2\}}_{\mathrm{M}} \rho a v_i = \rho u v_i \,,$$

so that the flux difference is given by

$$\frac{f^{\rho v}_{i+1/2} - f^{\rho v}_{i-1/2}}{\Delta x} = \frac{\rho u v_{i+1} - \rho u v_i}{\Delta x} = \frac{\rho u \Delta_{i+1,i} v}{\Delta x}, \tag{6.1}$$

which is a simple upwinding for the shear wave. Let v be a smooth solution. With the following Taylor expansion about grid node x_i

$$v(x_i + \Delta x) = v(x_i) + v_x(x_i)\Delta x + \tfrac{1}{2} v_{xx}(x_i)\Delta x^2 + \mathcal{O}(\Delta x^3),$$

the term on the RHS of (6.1) leads to the flux difference

$$\frac{\rho u \Delta_{i+1,i} v}{\Delta x} = \rho u (v_x + \tfrac{1}{2} v_{xx} \Delta x) = u\frac{\partial}{\partial x}\rho v + \frac{1}{2} u \frac{\partial^2}{\partial x^2}\rho v \Delta x + \mathcal{O}(\Delta x^2).$$

Using this, along with the results for $M > 0$, we obtain the modified equation for the transport of a shear wave:

$$\frac{\partial}{\partial t}\rho v + u\frac{\partial}{\partial x}\rho v = \frac{1}{2}|u|\frac{\partial^2}{\partial x^2}\rho v \Delta x + \mathcal{O}(\Delta x^2).$$

The numerical Reynolds number for the shear wave can now be calculated to

$$\boxed{\mathrm{Re}_{\mathrm{AUSM}} = \frac{|u|}{|u|\Delta x} = \frac{1}{\Delta x}}$$

Concerning the artificial viscosity on the shear wave, the AUSM scheme behaves like the Roe scheme. This is in agreement with the numerical results presented in the following sections. Interestingly the numerical Reynolds number for the entropy wave can be shown to have the order relation

$$\mathrm{Re}_{\mathrm{AUSM}} = \mathcal{O}_\mathrm{S}\Big(\frac{1}{M\Delta x}\Big). \tag{6.2}$$

6.4. Numerical results

For the numerical experiments presented in this sections we used the same test cases and settings as in the preceding sections, cf. Section 4.2.3 for a detailed description. The only difference had to be made for the AUSM scheme: for small Mach numbers the scheme becomes unstable for the standard setting of the CFL numbers $c = 0.5$ and had to be reduced to $c = 0.25$. This behaviour can be understood with the artificial viscosity on the entropy wave being $\mathcal{O}_\mathrm{S}(M\Delta x)$ or, equivalently, with the numerical Reynolds number given in Equation (6.2).

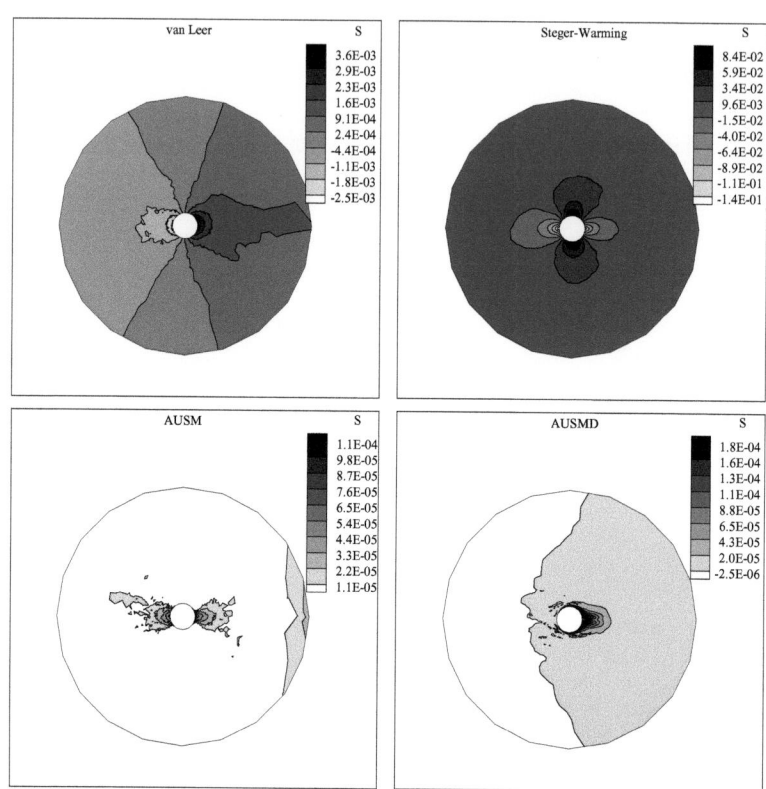

Figure 6.1.: Steady entropy distribution for the flow around a cylinder at $M_0 = 10^{-3}$, obtained with various first-order flux vector splittings. Shown are the contour lines of the entropy fluctuation $\tilde{s}$, cf. Equation (2.6), with different gray scales.

Entropy production and transport

In Figure 6.1 the contour lines of the entropy fluctuation, cf. Equation (2.6), for the steady flow around a cylinder are shown. A convective transport of the entropy by the flow can only be made out for the AUSMD scheme, which was added here to compare different schemes from the AUSM family. All flux vector splittings do not indicate a flow direction. The van Leer scheme has an entropy sink in the front stagnation point and an entropy source at the downwind stagnation point. The Steger-Warming scheme has two sources and two sinks of entropy. In both flux vector splittings the entropy is transported away from these points of entropy production by an isotropic process, i.e. by dissipation. The maximum entropy fluctuation $s_{\text{fluc}} = \tilde{s}_{\max} - \tilde{s}_{\min}$ is for AUSM and AUSMD given by

$$s_{\text{fluc}} \approx 10^{-4}$$

and

$$s_{\text{fluc}} \gg 10^{-4}$$

for the van Leer and Steger-Warming splittings.

Comparison with the analytical solution

In Figure 6.2 we present the isolines of pressure for the flow around a cylinder at an inflow Mach number of $M_{\text{in}} = 10^{-5}$. Top left there is the analytical solution of the incompressible potential flow and underneath there are the results obtained with AUSMD and AUSM. Top right is the approximate solution of Stokes flow; the results obtained with the van Leer and Steger-Warming are underneath.

AUSMD and AUSM have a pressure distribution similar to the incompressible potential flow solution shown above in the figure. Both schemes have the correct order of magnitude of the pressure variations. The AUSM scheme shows oscillations in the pressure field. The deviation from symmetry for AUSMD is due to the finite numerical Reynolds number that can never reach the physical Reynolds number for inviscid flow Re = ∞. As a consequence the flow in the numerical simulation separates from the cylinder and builds up a von Kármán vortex street where the average pressure is lower than in the upwind stagnation point.

Van Leer and Steger-Warming splittings produce a pressure field, which shows similarity with the pressure distribution of creeping flow.

Magnitude of pressure

In Figure 6.3 we present the maximal pressure fluctuation

$$p_{\text{fluc}} = \frac{p_{\max} - p_{\min}}{p_{\max}}$$

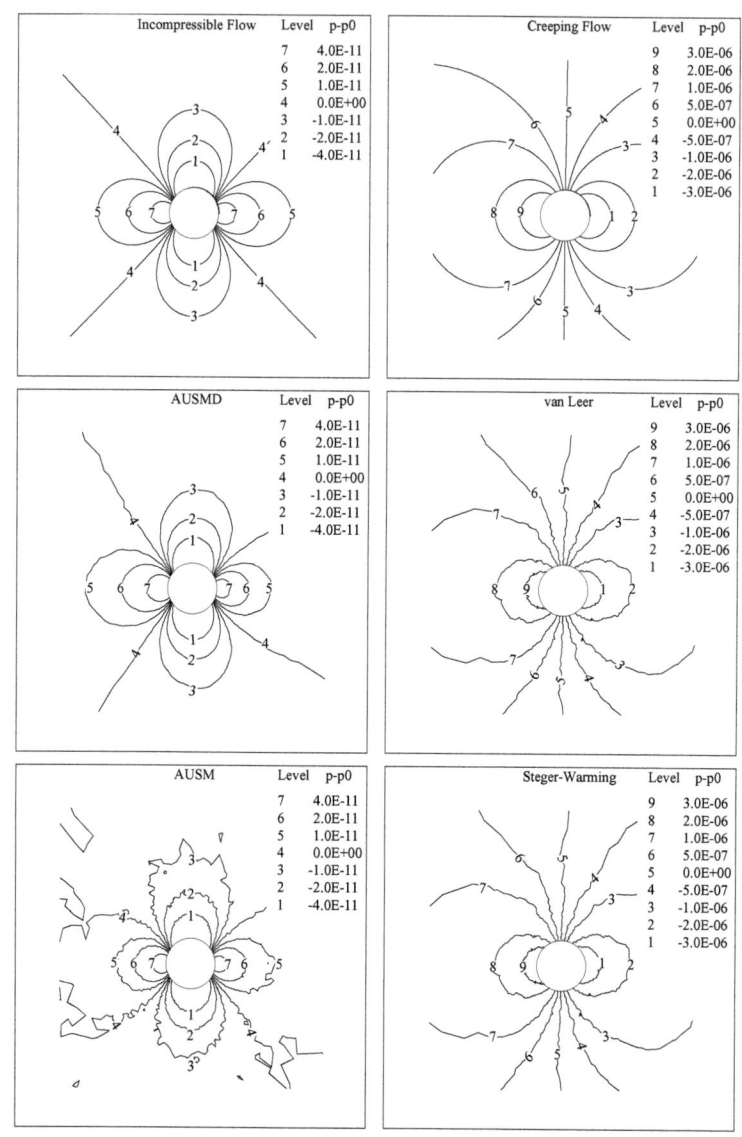

Figure 6.2.: Isolines of pressure for the flow around a cylinder at $M_0 = 10^{-5}$.
Left column: incompressible flow and asymptotically consistent schemes.
Right column: Creeping flow and asymptotically inconsistent schemes.

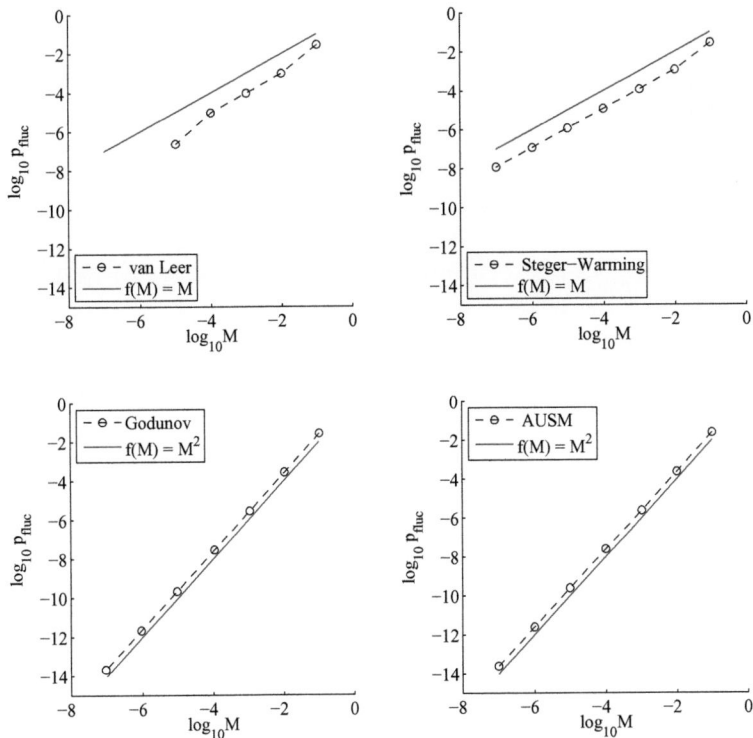

Figure 6.3.: Maximal pressure fluctuation $p_{\text{fluc}} = (p_{\max} - p_{\min})/p_{\max}$ against inflow Mach number for the flow around a cylinder obtained with various explicit first-order upwind schemes.

obtained with various flux vector splittings and, for comparison, the results obtained with the exact Riemann solver, i.e. the Godunov scheme. We can see, that the Steger-Warming and the van Leer splitting have a pressure fluctuation of the order of magnitude $\mathcal{O}_S(M)$. A plausible explanation for this behaviour was given with the analogy to creeping flow. On the other hand, the AUSM splitting shows a pressure fluctuation of the correct order relation

$$p_{\text{fluc}} = \mathcal{O}_S(M^2) \,.$$

6.5. Summary

We summarise the results using the concept of *asymptotic consistency*:

Theorem 6.5.1. *The following flux vector splittings are **asymptotically inconsistent** with respect to the shear wave:*

- *The van Leer splitting (cf. Section 6.1)*
- *The Steger-Warming splitting (cf. Section 6.2)*

*The Liou-Steffen scheme (AUSM) is **asymptotically consistent** (cf. Section 6.3) with respect to the shear wave.*

Proof. The proofs can be found in the corresponding sections. □

Since a characteristic decomposition is not possible, the AUSM scheme could not be tested for the dissipative treatment of the acoustic waves. Nevertheless, the numerical results justify its characterisation as *asymptotical consistent*, even though the CFL number had to be slightly decreased.

7. Pressure decomposition

In the *continuous* asymptotic analysis of the low Mach number Euler equations, cf. Section 2.3.2, the (scaled) pressure field is expanded in a asymptotic 3-term series with the *asymptotic sequence* $\phi_n(M) = M^n$, $n = 0, 1, \ldots$

$$p = p^{(0)} + M p^{(1)} + M^2 p^{(2)} + o(M^2) \quad \text{as} \quad M \to 0.$$

In the absence of acoustics the one-scale asymptotic analysis reveals that $p^{(1)}$ is constant and can be subsumed in $p^{(0)}$ – the pressure fluctuations are then $\mathcal{O}_S(M^2)$. We now look for a heuristic explanation for the appearance of a pressure of order $\mathcal{O}_S(M)$ in some numerical results. This is done with a heuristic argument relying on a simple order analysis.

7.1. Viscosity-Induced Pressure Fields

For a pressure decomposition in terms of the two small parameters, the Mach number M and the grid size Δx, physical *and* numerical effects have to be taken into account. Let us begin with an analogy to *physical viscosity*. Viscous forces in non-trivial flows lead to a pressure fluctuation of the order

$$\tilde{p}_{\text{visc}} = \mathcal{O}_S\left(\eta \frac{u_{\text{ref}}}{l_{\text{ref}}}\right),$$

where η is the coefficient of dynamic viscosity and $u_{\text{ref}}/l_{\text{ref}}$ gives the order of magnitude of the shearing in the flow. Note that we exclude trivial (or uniform) flows with the velocity being free of shear layers, in which case the order relation is no longer valid (no pressure fluctuation at all). We express η by the Reynolds number

$$\text{Re} = \frac{\rho_{\text{ref}} l_{\text{ref}} u_{\text{ref}}}{\eta}$$

and obtain an order relation for the viscosity-induced pressure fluctuations

$$\tilde{p}_{\text{visc}} = \mathcal{O}_S\left(\frac{M^2}{\text{Re}}\right).$$

If the viscosity has its origin in the numerical dissipation of shear layers, the physical Reynolds number Re has to be replaced by its numerical counterpart Re_{num}, and the relation for the numerically induced pressure component becomes

$$p_{\text{num}} = \mathcal{O}_S\left(\frac{M^2}{\text{Re}_{\text{num}}}\right). \tag{7.1}$$

Equation (7.1) will allow us to find the order relation for the pressure variations first-order upwind schemes produce in the low Mach number regime.

7.1.1. Asymptotically consistent schemes

Schemes that are *asymptotically consistent* with respect to the shear wave have a corresponding numerical Reynolds number satisfying

$$\text{Re}_{\text{num}} = \mathcal{O}_S\left(\frac{1}{\Delta x}\right),$$

so that the numerically induced pressure field can be derived from (7.1) as

$$p_{\text{num}} = \mathcal{O}_S(\text{M}^2 \Delta x).$$

A sensible decomposition of the pressure would therefore be

$$p = p^{(0)} + \text{M}^2 p^{(2)} + \text{M}^2 \Delta x\, p_{\text{num}}^{(2)} + o(\text{M}^2) \quad \text{as} \quad \text{M} \to 0, \tag{7.2}$$

It consists of a constant background pressure $p^{(0)}$, a pressure variation $\text{M}^2 p^{(2)}$ induced by Bernoulli's principle and the pressure variation $\text{M}^2 \Delta x\, p_{\text{num}}^{(2)}$ induced by the numerical viscosity of shear layers, which is also of the order of the square of the Mach number and, at the same time, of the order of the grid cell size. The Decomposition (7.2) agrees with the observations made in our numerical simulations in several aspects:

1. Even for coarse grids, the cell size Δx is always a small fraction of the computational domain, so that we conclude

    $$\Delta x \ll 1 \quad \Rightarrow \quad \text{M}^2 \Delta x\, p_{\text{num}}^{(2)} \ll \text{M}^2 p^{(2)}.$$

    This explains, why a numerically induced pressure field cannot be observed. Especially the observed maximal pressure fluctuation

    $$p_{\text{fluc}} = \frac{p_{\max} - p_{\min}}{p_{\max}}$$

    is independent of the cell size Δx as can be seen in Figure 7.1 and 7.2.

2. *Asymptotically consistent* schemes do not produce a pressure field of the wrong order $\mathcal{O}_S(\text{M})$ as can be verified for the Roe scheme in Figure 4.6 and for the AUSM scheme in Figure 6.3.

3. For the flow around a cylinder, the maximal (numerical) pressure fluctuation p_{fluc} agrees with the theoretical value Δp_{pot} for incompressible potential flow:

    $$p_{\text{fluc}} \approx \Delta p_{\text{pot}} = 2\gamma \text{M}^2. \tag{7.3}$$

    In Figures 7.1 and 7.2 the line $p = 2\gamma \text{M}^2$ is shown with a dashed line.

7.1.2. Asymptotically inconsistent schemes

The class of schemes which are *asymptotically inconsistent* with respect to the shear wave have a numerical Reynolds number for this characteristic wave satisfying

$$\text{Re}_{\text{num}} = \mathcal{O}_S\Big(\frac{M}{\Delta x}\Big),$$

so that the order relation for the numerically induced pressure field is according to (7.1):

$$p_{\text{num}} = \mathcal{O}_S(M\Delta x).$$

In the absence of acoustic waves the pressure field can be decomposed as

$$p = p^{(0)} + M^2 p^{(2)} + M\Delta x\, p^{(1)}_{\text{num}} + o(M^2) \quad \text{as} \quad M \to 0, \tag{7.4}$$

where $p^{(0)}$ is the constant background pressure, $M^2 p^{(2)}$ is the physically induced pressure variation according to Bernoulli's principle, and $M\Delta x p^{(1)}_{\text{num}}$ is a numerically induced pressure field. Decomposition (7.4) is also in agreement with our numerical observations in the following aspects:

1. For small Mach numbers the numerically induced pressure exceeds the physical pressure variations:

$$M \ll \Delta x \quad \Rightarrow \quad M\Delta x\, p^{(1)}_{\text{num}} \gg M^2 p^{(2)}.$$

   The flow degenerates – due to numerical viscosity – to a flow type similar to *creeping flow* with its characteristic pressure field as shown in Figure 4.3 (left).

2. On a fixed mesh, the numerically induced pressure field is of the order $\mathcal{O}_S(M)$ as can be seen in Figure 5.4.

3. For a fixed Mach number, the numerically induced pressure field is proportional to the cell size Δx as can be seen in Figure 7.1 and 7.2 for an inflow Mach number of $M_\infty = 10^{-2}$ and $M_\infty = 10^{-7}$, respectively.

7.2. Numerical results

All calculations were done with the standard settings presented in Section 4.2.3 for the flow around a cylinder with 9800 grid cells. The difference between the calculations that lead to the diagrams in Figure 7.1 and 7.2 is the inflow Mach number: $M_\infty = 10^{-2}$ for Figure 7.1 and $M_\infty = 10^{-6}$ for Figure 7.2. The similarity between the diagrams in both figures shows that the relation between pressure fluctuation and cell size is independent of the chosen Mach number. Note however, the difference between the absolute value of the pressure fluctuations indicated by the scaling of the vertical axes of the diagrams. Also note that all y-axes have a linear and not a logarithmic scale. The horizontal axis has ticks representing the nth refinement of the original grid with

270 grid cells. A refinement of the irregular mesh is done by decomposing a triangle into four parts so that the edges of the new grid has approximately half the length of the predecessor:

$$\Delta x_{n+1} \approx \frac{1}{2} \Delta x_n .$$

We point out that in our irregular grid the cells have triangular shape with varying edge lengths, so that the above statement is only true in an approximate sense, i.e. Δx_n denotes the average edge length of all triangles in the grid after the nth refinement.

The *asymptotically inconsistent* schemes show a slight deviation from a linear relation between pressure fluctuation p_{fluc} and cell size Δx, especially for the coarsest grid. The reason therefore might lie in the effect of second-order errors which mix for larger Δx with the first-order errors. The results obtained with the *asymptotically consistent* schemes, presented on the right column of Figure 7.1 and 7.1, approximate the pressure fluctuation of incompressible potential flow very well.

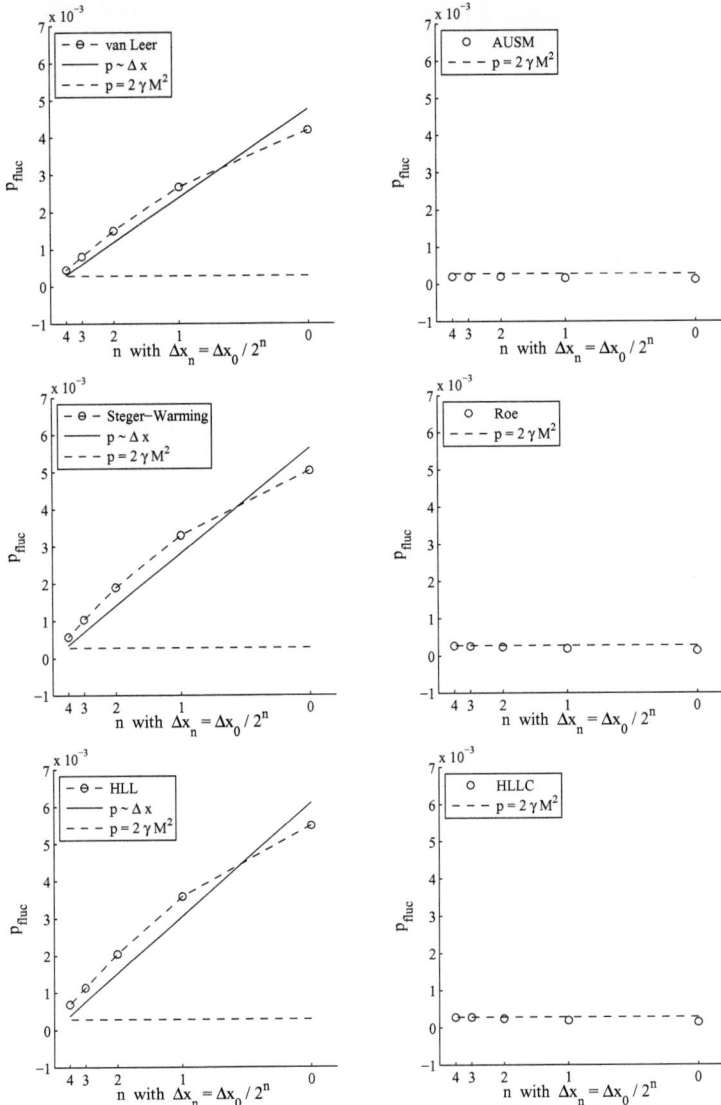

Figure 7.1.: Maximal pressure fluctuation $p_{\text{fluc}} = (p_{\max} - p_{\min})/p_{\max}$ against cell size $\Delta x_n = \Delta x_0/2^n$ for the flow around a cylinder at $M_\infty = 10^{-2}$ for various upwind schemes.

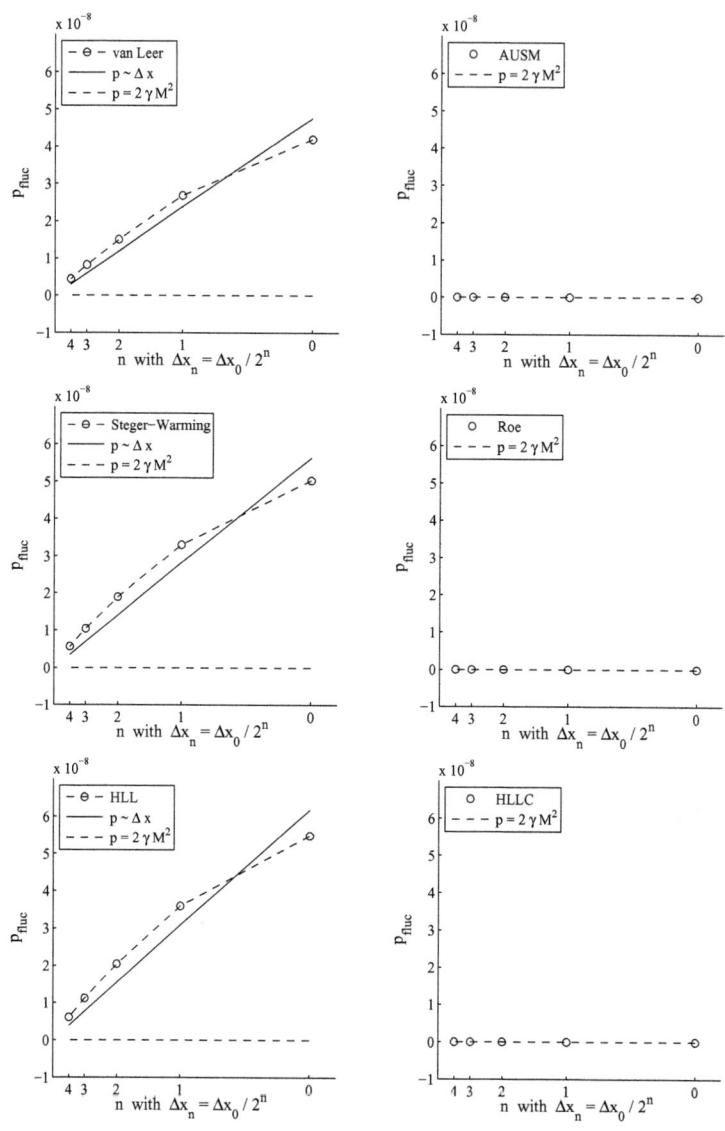

Figure 7.2.: Pressure fluctuation $p_{\text{fluc}} = (p_{\max} - p_{\min})/p_{\max}$ against cell size for the flow around a cylinder at $M_\infty = 10^{-7}$ for various upwind schemes.

Part II.
Analysis for 2D: cell geometry

8. Numerical experiments

In this chapter we present a variety of numerical experiments that all suggest an intriguing result: the *accuracy problem* for the first-order Roe scheme depends on the geometry of the finite volume cell. On Cartesian grids the accuracy of the results deteriorates with decreasing Mach numbers, while the accuracy remains unchanged if the finite volume cell is of triangular shape.

All calculations and grids presented in this chapter were done with software developed by the author, cf. Appendix A.6 for more information. The discretisation uses the first-order Roe scheme in conjunction with the explicit forward Euler time discretisation.

8.1. Flow around a cylinder

The flow around a cylinder[1] is very useful since we know the incompressible potential flow solution, i.e. we have an analytical reference solution. It can be shown by asymptotic analysis that the deviation in the pressure fluctuation between low Mach number and incompressible flow is $\mathcal{O}(\mathrm{M}^4)$, which justifies this comparison.

The flow domain Ω is restricted by the radius of the cylinder r_0 and the radius of the circular computational domain r_1

$$\Omega = [r_0, r_1] \times [\phi_0, \phi_1] = [0.5, 5.0] \times [0, 2\pi] \, .$$

For the body-fitted, structured grid we use a division into $n_\phi = 150$ cells along the circumference and $n_r = 50$ along the radius. For the unstructured, body-fitted grid we use 9800 triangular cells.

The *initial conditions* are uniform

$$\rho_0 = 1.0 \, ,$$
$$\mathbf{u}_0 = (u_0, 0)^T \, ,$$
$$p_0 = 1.0 \, ,$$

where the absolute value $\|\mathbf{u}_0\| = u_0$ of the initial velocity is set to meet the prescribed initial Mach number

$$\mathrm{M}_0 = \frac{\|\mathbf{u}_0\|}{a_0} = \frac{\|\mathbf{u}_0\|}{\sqrt{\gamma p_0/\rho_0}} = \frac{\|\mathbf{u}_0\|}{\sqrt{\gamma}} \, ,$$

---

[1] The correct term would be *infinite cylinder*. Since in this case the solution does not depend on the symmetry axis it is legitimate to reduce the problem to two dimensions.

where γ is the adiabatic index and set to 1.4 throughout our calculations. The exact solution at infinity is assumed uniform

$$\rho_\infty = 1.0,$$
$$\mathbf{u}_\infty = (u_\infty, 0)^T,$$
$$p_\infty = 1.0.$$

In the *far-field boundary conditions* this solution is assumed to be a good approximation of the solution at the outer radius r_1. Therefore the computational domain is embedded by a layer of ghost cells in which we impose the solution at infinity:

$$\rho_{\text{ghost}} = 1.0,$$
$$\mathbf{u}_{\text{ghost}} = (u_\infty, 0)^T,$$
$$p_{\text{ghost}} = 1.0.$$

This is a good approximation whenever the external boundary is far from obstacles inside the domain. The upwind-character of the scheme automatically takes into account, whether the flow is entering (inflow) or leaving the domain (outflow) at an interface on the boundary. Note that in most calculations shown in this thesis initial conditions and the conditions at infinity are set equal with the consequences, that

$$\text{M}_\infty = \text{M}_0 \quad \text{and} \quad \mathbf{u}_\infty = \mathbf{u}_0.$$

For this reason we refer to far-field, inflow or initial Mach number with the same symbol M_0 if not stated otherwise.

8.1.1. Grid transformation: from rectangular to triangular cells

We start with a regular cylinder grid, continuously transform it to a triangular cylinder grid and observe the numerical behaviour of the scheme. This transition is implemented by reducing the original length Δx_0 of one of the four edges of the quasi square cell by a factor 10^n. The resulting edge has length

$$\Delta x = \frac{1}{10^n} \Delta x_0,$$

and belongs to a trapezoid, which converges for $n \to \infty$ to a triangular cell. In the finite volume context the flux across the "squeezed" edge also converges to zero. In Figure 8.1, on the left column, a detail of the grid shows the cell shape used in the simulation presented on the right. On top is a detail of the original regular grid, in the middle the squeezed cell edge has only one fourth of the original length and in the bottom figure it is reduced to a hundredth. The corresponding pressure fields are presented on the right column. On the regular grid (top) the solution is completely wrong. The trapezoidal cells in the middle lead to a solution that is "on the right way" but still far from the characteristic quadrupole character (cf. Figure 2.2) and the pressure variations are still to large. The isolines of pressure in the bottom figure suggest good accuracy. Further reduction of Δx does not improve the accuracy – the ratio has reached a critical value. The question is, does the critical ratio depend on the inflow Mach number?

In Figure 8.2 we show the results of a study to clear this question. It can be seen, how the accuracy improves, when the cell shape approaches a triangle. The abscissa shows the ratio of original edge length Δx_0 to the reduced edge length Δx on a logarithmic scale:

$$\log \frac{\Delta x}{\Delta x_0} \leadsto \text{abscissa}.$$

The ordinate represents the pressure deviation from the incompressible potential flow solution on a logarithmic scale,

$$\|\Delta p\|_2 = \log \sqrt{\sum_i^{\#\text{cells}} (p_i^{\text{Roe}} - p_i^{\text{potFlow}})^2} \leadsto \text{ordinate}.$$

Comparing the diagrams, we see that for $M_0 = 10^{-1}$ the L_2-error of the pressure stagnates if the edge length Δx is reduced by a factor of about 10^1, while for $M_0 = 10^{-2}$ the stagnation begins for a length reduction larger than 10^2 and so on. In other words, the accuracy problem is avoided on trapezoidal finite volume cells if the lengths of the parallel edges Δx and Δx_0 satisfy

$$\frac{\Delta x}{\Delta x_0} = \mathcal{O}_S(M).$$

Unfortunately, we have no analytical explanation of this behaviour so far.

8.1.2. Structured vs. unstructured triangular grids

The previous experiment on *structured grids* leaves an open question: is the *accuracy problem* present on unstructured triangular grids? To find an answer we compare the flow around a cylinder calculated on a structured and an unstructured grid. Both grids shown in Figure 8.3 are body-fitted with some 10.000 triangular finite-volume cells. The isolines of pressure, shown on the right column of Figure 8.3, are of comparable accuracy on structured and unstructured grids. The same holds for the maximum of the pressure fluctuation (which is not shown in the figure).

The unstructured grid has slightly higher resolution near the cylinder surface but much coarser grids lead to the same conclusion: the *accuracy problem* does not occur on triangular finite volume cells – independent of the grid structure.

Note on unstructured quadrilateral grids

A similar experiment, not presented here, was done with an unstructured grid of quadrilateral finite volume cells. The deviation of the regular grid was obtained with

Figure 8.1.: Flow around a cylinder with the first-order Roe scheme on a grid with 150x50 cells at $M_0 = 10^{-2}$. Left column: trapezoidal cells converging to triangular cells. Right column: corresponding isolines of pressure converging to potential flow solution. The pressure maximum $p_{max} - 1$ decreases from $4.2 \cdot 10^{-4}$ (top) over $1.6 \cdot 10^{-4}$ (middle) down to $7.7 \cdot 10^{-5}$ (bottom), which is close to the potential flow value of $5.0 \cdot 10^{-5}$.

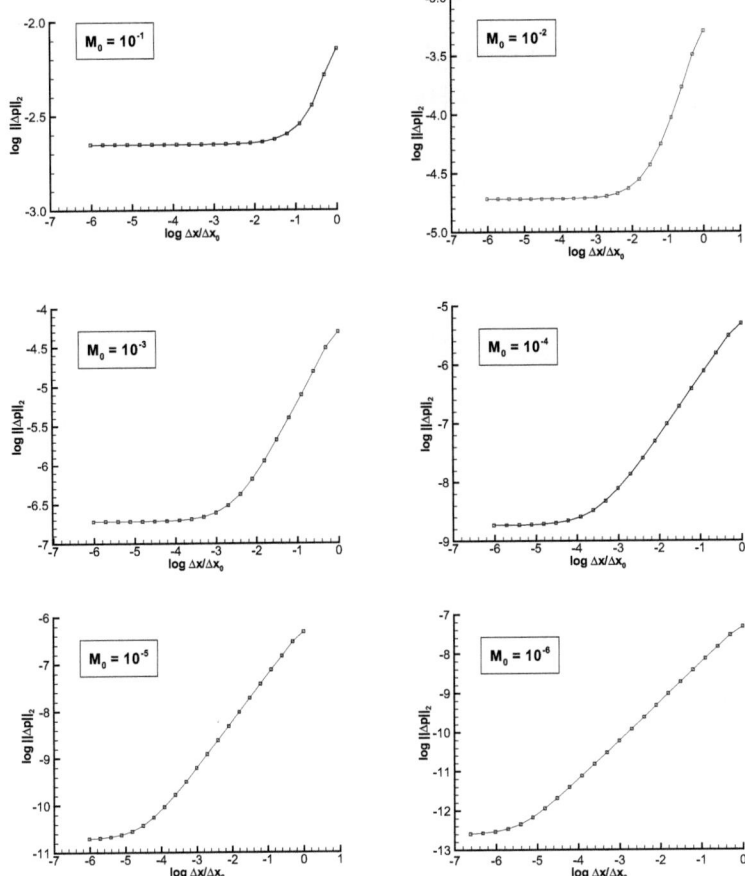

Figure 8.2.: L_2-error of the pressure as a function of grid cell shape: $\Delta x = \Delta x_0$ represents regular grid cells and $\Delta x/\Delta x_0 \approx 0$ approximately triangular grid cells. Results obtained for the flow around a cylinder with the explicit first-order Roe scheme on a structured grid (cf. Figure 8.3) with $n_\phi \times n_r = 150 \times 50$ cells at $M_0 = 10^{-1}$ down to $M_0 = 10^{-6}$.

a random shift of the grid vertices. The isolines of pressure, compared to the regular grid, did not improve. We therefore conclude, that the reason for the *accuracy problem* is not linked to the grid structure as a whole, but to the cell geometry.

8.1.3. A layer of perturbation cells

The negative effect of a layer of trapezoidal finite volume cells surrounded by triangular cells can be seen in Figure 8.4. On the left, a detail of the structured grid with 150 angular and 50 radial cells is shown. The contour lines of the pressure, shown on the right, deviate from the physically correct solution in the vicinity of the quadrilateral cells. There, the erroneous pressure variations are $\mathcal{O}_S(M)$ and make the physical $\mathcal{O}_S(M^2)$ pressure field invisible for decreasing Mach numbers.

8.1.4. Convergence study

By construction, Roe's upwind scheme without reconstruction is first-order accurate in space. But how is this order of accuracy obtained on grids with triangular compared to quadrilateral finite volume cells? In Figure 8.5 we present a convergence study for the flow around a cylinder with the explicit first-order Roe scheme. The abscissa is associated with the angular number of grid cells n_ϕ, i.e. the number of cells along the circumference of the cylinder. The different refinement steps are given in terms of $n_\phi \times n_r$, with n_r being the number of grid cells in radial direction:

- $30 \times 10 = 300$,
- $60 \times 20 = 1200$,
- $120 \times 40 = 4800$,
- $240 \times 80 = 19\,200$,
- $480 \times 160 = 76\,800$.

The relation to the average size of the grid cells in radial $\bar{\delta}_r$ and in angular $\bar{\delta}_\phi$ direction is given by

$$\bar{\delta}_r \approx \frac{r_1 - r_0}{n_r} = \frac{4.5}{n_r},$$
$$\bar{\delta}_\phi \approx \frac{2\pi \bar{r}}{n_\phi} \approx \frac{4\pi}{n_\phi},$$

where the radius of the flow domain is $r_1 = 5$ and the radius of the cylinder is $r_0 = 0.5$. The grid was generated in a way to assure the optimal ratio $\delta_\phi/\delta_r \approx 1$ throughout the grid.

The ordinate axis is associated with the L_2-error of the pressure on a logarithmic scale, where the incompressible potential flow served as reference solution. The curves

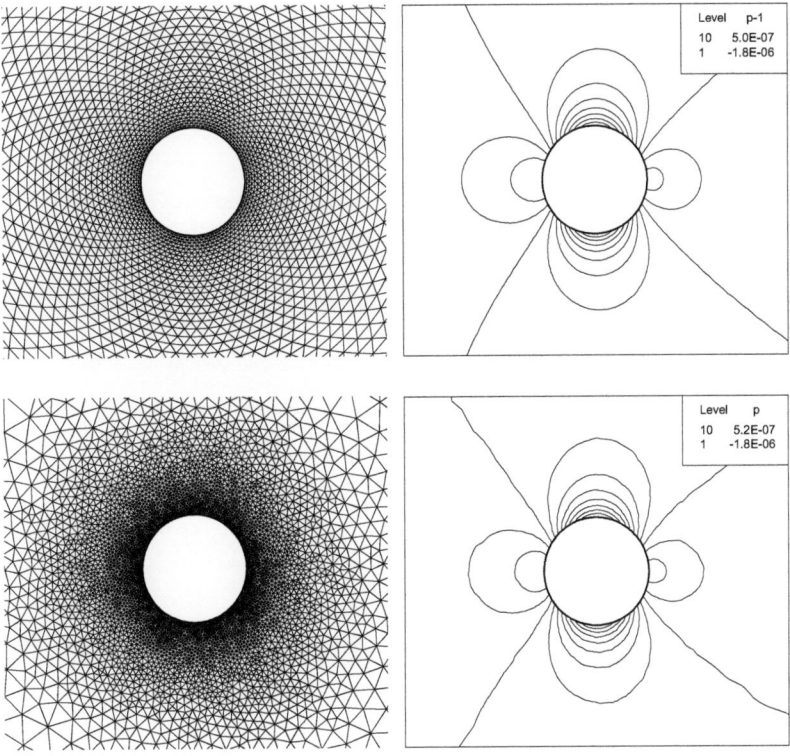

Figure 8.3.: Flow around a cylinder with the first-order Roe scheme at $M_0 = 10^{-3}$. Left column: structured, body-fitted grid with 150×50 triangular cells (top). Unstructured, body-fitted grid with 9800 triangular cells (bottom). Right column: isolines of pressure obtained on the corresponding grids.

 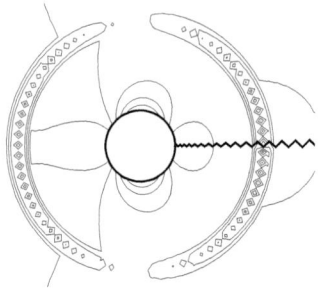

Figure 8.4.: Left: detail of the structured triangulation around a disc with a layer of quadrilateral cells. Right: isolines of pressure for the flow around a cylinder with the first-order Roe scheme on 150x50 cells at $M_0 = 10^{-3}$.

are straight lines with slopes indicating first-order accuracy for both cell types. Yet, the pressure error and the pressure fluctuations on a fixed *quadrilateral grid* are $\mathcal{O}_S(M)$ instead of $\mathcal{O}_S(M^2)$. This has negative consequences for numerical simulations: take the case with an inflow Mach number of $M_0 = 10^{-1}$. To obtain the same accuracy on quadrilateral as on triangular cells, the number of cells has to be increased from 120×40 (triangles) to 480×160 (quadrilaterals), i.e. 16 times more cells are needed. This discrepancy worsens for lower Mach numbers.

8.1.5. Efficiency study

There is no doubt, that phenomena on the time scale of the flow, such as the transport of entropy, humidity or a chemical substance are practically frozen with an explicit upwind scheme in low Mach number flow, since the number of time steps needed increases like $\mathcal{O}_S(1/M)$. On the other hand, acoustic phenomena are calculated in $\mathcal{O}_S(1)$ time steps. Interestingly, the adaption of the flow field to a perturbation such as a moving obstacle happens on the acoustic time scale. The following numerical experiment serves to measure the time from the sudden onset of fluid motion (uniform initial conditions) to the steady state. Once the acoustic waves have left the domain, a steady pressure field settles around the cylinder. The passed CPU time (Pentium 4 with 3.0 GHz) till convergence is presented in the left of Figure 8.6. On the right, a similar diagram shows the number of time steps needed. Interestingly, the time for the pressure field to settle down is of the order of the logarithm of the Mach number:

$$t_{\text{CPU}} = \mathcal{O}(\log M_0) .$$

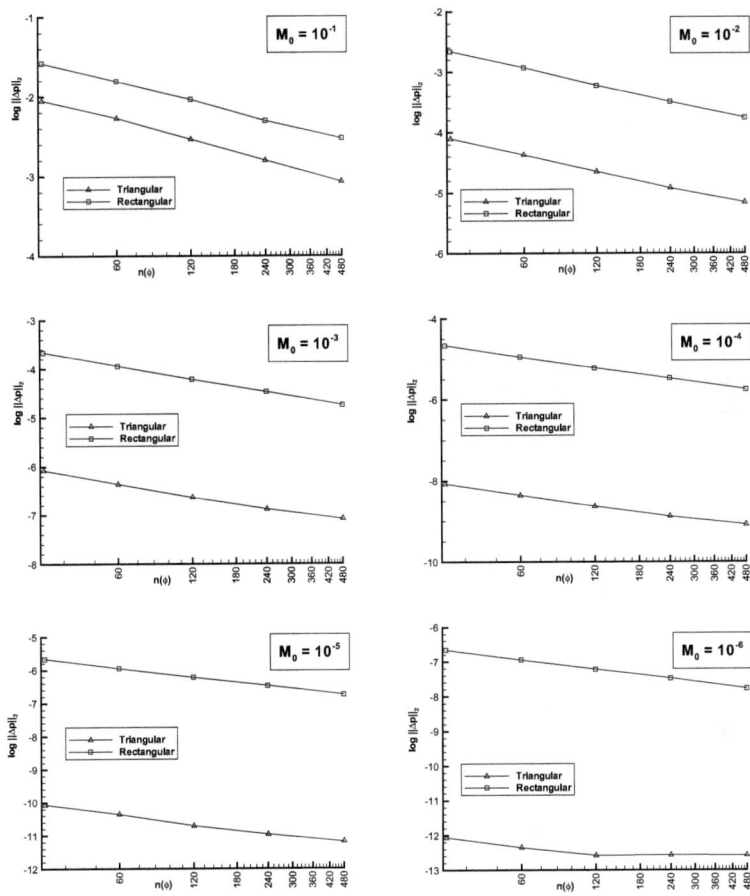

Figure 8.5.: L_2-error of the pressure as a function of the grid refinement for the flow around a cylinder with the first-order Roe scheme at $M_0 = 10^{-1}$ down to $M_0 = 10^{-6}$. The total number of grid cells in the structured and body-fitted grid are: $n_\phi \times n_r = 30 \times 10$, 60×20, 120×40, 240×80 and 480×160, where n_ϕ is the number of cells around the cylinder and n_r from the cylinder surface to the far-field boundary.

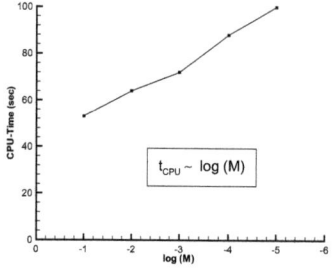 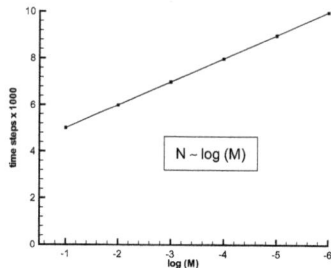

Figure 8.6.: Efficiency study for the flow around a cylinder (150x50 triangular cells) with the explicit first-order Roe scheme on a Pentium 4 with 3.0 GHz. CPU time (left) and number of explicit time steps (right) for convergence to steady state from homogeneous initial conditions. The pressure field settles to steady state after $\mathcal{O}(\log(M))$ time steps.

For the class of low Mach number flows around obstacles, such as aerofoils, a standard upwind scheme on primary grids with triangular cells is rather efficient. We do not have similar studies for preconditioned implicit methods, but it would be interesting to see, whether these newly developed schemes are competitive.

8.2. Flow around a square

The flow around a square has two advantages: first we can work on a Cartesian grid, which simplifies the later analysis and secondly the behaviour of the solution suggests the origin of the numerical error. We compare the solution obtained on a Cartesian grid with the one obtained on a related triangulation.

Triangular grid derived from a Cartesian grid

The triangulation we consider in the following experiments is derived from a Cartesian grid by introducing diagonals into the square cells as additional edges. The construction principle can be seen in Figure 8.7. Note that the diagonals are introduced in a uniform direction from the lower left to the upper right corner of the Cartesian cells. The resulting grids for the flow around a square are depicted on the left column of Figure 8.8. To make the calculations on the flow domain

$$\Omega = [0.4, 0.6] \times [0.4, 0.6] \,/\, [0, 1] \times [0, 1]$$

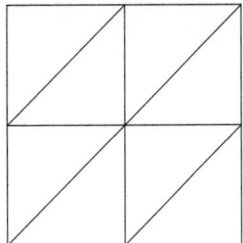

Figure 8.7.: Triangular grid originating from a Cartesian grid by introducing the diagonal running from the lower left to the upper right corner of the square cell.

comparable, the total number of cells were chosen to be similar: $2 \times 35 \times 35 = 2450$ in the triangulation and 50×50 in the Cartesian mesh (minus the cells covered by the rectangular obstacle).

The initial and boundary conditions are set similar to the ones for the flow around a cylinder. Recall, the *initial conditions* are uniform

$$\rho_0 = 1.0 \,,$$
$$\mathbf{u}_0 = (u_0, v_0)^T \,,$$
$$p_0 = 1.0 \,,$$

where the absolute value $\|\mathbf{u}_0\|^2 = u_0^2 + v_0^2$ of the initial velocity is set to meet the prescribed initial Mach number

$$\mathrm{M}_0 = \frac{\|\mathbf{u}_0\|}{a_0} = \frac{\|\mathbf{u}_0\|}{\sqrt{\gamma p_0/\rho_0}} = \frac{\|\mathbf{u}_0\|}{\sqrt{\gamma}} \,.$$

The computational domain is embedded by a layer of ghost cells in which we impose the solution at infinity:

$$\rho_{\text{ghost}} = 1.0 \,,$$
$$\mathbf{u}_{\text{ghost}} = (u_\infty, v_\infty)^T \,,$$
$$p_{\text{ghost}} = 1.0 \,.$$

The angle of attack α is then given by

$$\tan \alpha = \frac{v_\infty}{u_\infty} \,.$$

Again initial and far-field condition are set according to the solution at infinity and the corresponding Mach numbers are simply referred to as M_0.

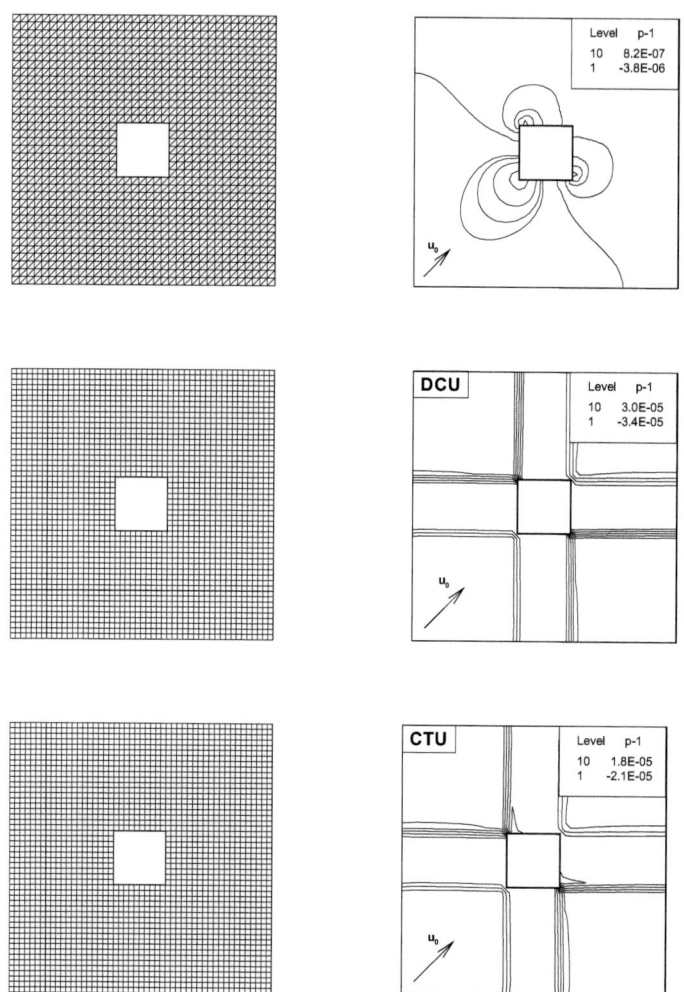

Figure 8.8.: Flow around a square with 45° angle of attack at $M_0 = 10^{-3}$ with the explicit first-order Roe scheme. Top: grid with $2 \times 35 \times 35$ triangular cells (left) and isolines of pressure. Middle/Bottom: grid with 50×50 rectangular cells (left) and isolines of pressure obtained with the donor cell upwind method with CFL = 0.45 (middle) and corner upwind transport method at CFL = 0.9 (bottom).

Randall Leveque suggested [33] to distinguish between the donor cell upwind method (DCU) and the corner transport upwind method (CTU), cf. the textbook by Leveque [34], in conjuction with the *accuracy problem*. Both methods are implemented with Roe's approximate Riemann solver for the normal and transverse fluctuations.

On the right of Figure 8.8 the isolines of pressure for an inflow Mach number of $M_0 = 10^{-3}$ and an angle of attack of $\alpha = 45°$ are shown. For this test case we do not know the exact solution, yet, we expect the maximum pressure in the stagnation point to be about $\frac{1}{2}\gamma M_0^2$ (Bernoulli's principle). This is well approximated on the triangular-cell grid and overestimated by a factor of 100 on the Cartesian grid, where, in addition, the isolines of pressure are grid-aligned. This unphysical behaviour for low Mach number flow can be observed for both – the donor cell upwind method and the corner transport upwind method [34]. In the latter method, the information not only from the neighbouring cells is used for a single cell update, as for the DCU, but also from the cells beyond the corners.

8.3. Low Mach number region

In some high speed applications the local Mach number

$$M_{local} = \frac{u_{local}}{a_{local}}$$

can drop below 0.3, so that small regions can be governed by low Mach number flow. A typical example is the stagnation point region at the front of an aerofoil. The question arising is, whether simulations on quadrilateral finite volume cells deviate from those executed on triangular cells.

We choose sonic inflow with $M_{inflow} = 1.0$ against the face of a square. This test case has a low Mach number region that is rather large, so that the effects are expected to be especially pronounced. In Figure 8.9 contours of density (left column) and pressure (right column) are shown for grids with rectangular (bottom line) and triangular finite volume cells (top line). Density and pressure contours look similar on triangular cells, as they should from theory, since they a coupled by the adiabatic relation

$$\rho \sim p^{1/\gamma}.$$

The results obtained on rectangular grid cells show an unphysical decoupling of density and pressure in the vicinity of the stagnation point.

8.4. Gresho vortex

An example of unsteady flow is a moving vortex as proposed by Gresho *et al.* in [18, 17]. This test case is run on two different grids:

- A Cartesian grid with 70×70 finite volume cells (cf. right of Figure 8.11)

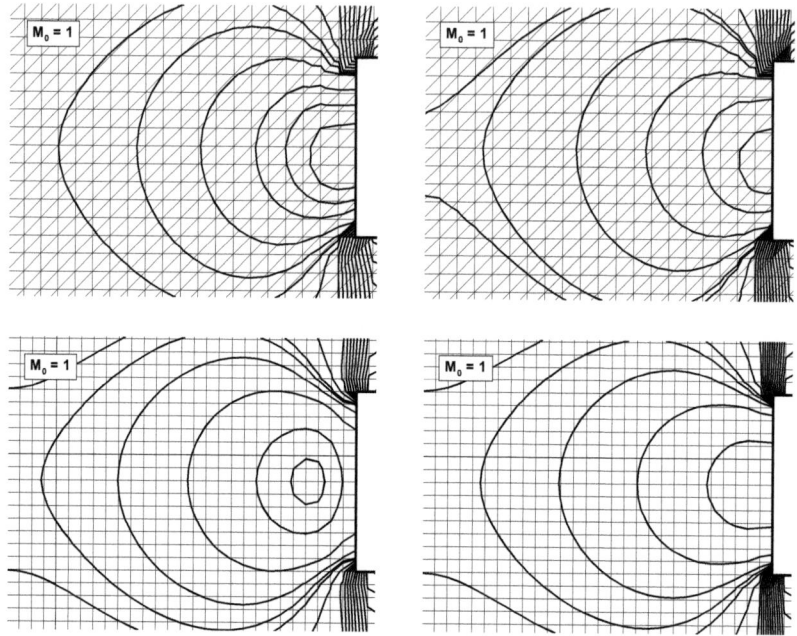

Figure 8.9.: Stagnation point of the flow around a square with $\alpha = 0°$ angle of attack at $M_{inflow} = 1.0$ with the explicit first-order Roe scheme. Left column: density contours for the grid with triangular cells (top) and square cells (bottom). Right column: pressure contours. In the low Mach number region the density decouples from the pressure field if calculated on rectangular cells.

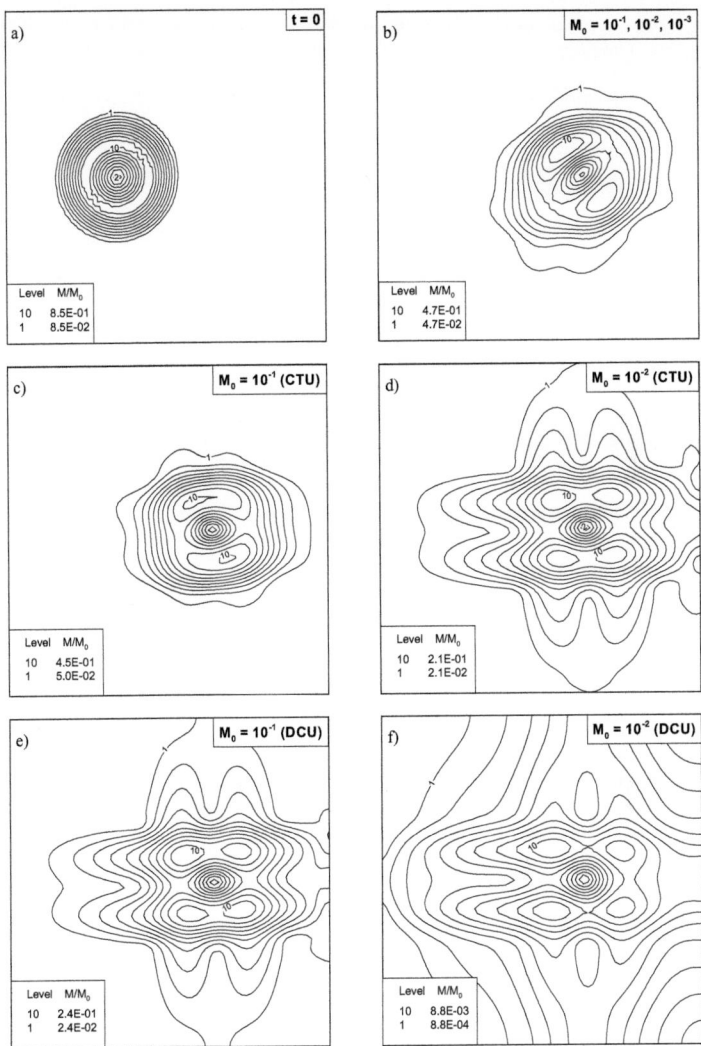

Figure 8.10.: Mach number contour plots for simulations of a Gresho vortex with the explicit first-order Roe scheme. a) Initial condition. b) Simulation on $2 \times 50 \times 50$ triangular cells at $M_0 = 10^{-1}$ down to $M_0 = 10^{-3}$. c) and d) CTU method on 70×70 Cartesian grid cells at $M_0 = 10^{-1}$ (left) and $M_0 = 10^{-2}$ (right). e) and f) DCU with the same settings.

- Triangulation, derived from a Cartesian grid as depicted in Figure 8.7, with $2 \times 50 \times 50$ triangular finite volume cells (cf. left of Figure 8.11)

At the initial time $t = 0$ a vortex of radius $R = 0.2$ at the position $(x_0, y_0) = (0.35, 0.5)$ is prescribed. Its radial velocity is given by

$$u_r(r) = u_0 \begin{cases} 2r/R & \text{if } 0 \leq r < R/2, \\ 2(1 - r/R) & \text{if } R/2 \leq r < R, \\ 0 & \text{if } R \leq r. \end{cases}$$

The vortex is superimposed to an initial, uniform background flow with

$$\rho_0 = 1.0,$$
$$\mathbf{u}_0 = (u_0, 0)^T,$$
$$p_0 = 1.0,$$

where the horizontal velocity is linked to the initial Mach number by

$$u_0 = \sqrt{\gamma} M_0.$$

Note that background velocity and rotational velocity of the Gresho vortex are linked by the same value u_0. Also note that the correct pressure field is known from the analytic solution. Nevertheless, it is left here to the scheme to set up the right pressure and density field in the process of calculation.

The *far-field boundary conditions* are implemented by imposing

$$\rho_{\text{ghost}} = 1.0,$$
$$\mathbf{u}_{\text{ghost}} = (u_\infty, 0)^T,$$
$$p_{\text{ghost}} = 1.0$$

in the ghost cells, where $u_\infty = u_0$, as for the cylinder flow. The test case is run for the Mach numbers $M_0 = 10^{-1}$, 10^{-2} and 10^{-3}.

The focus of this experiment is not the comparison with the exact solution, as the absolute accuracy is expectedly poor for a first-order scheme on such a coarse grid. We are rather interested in the dependency of the accuracy on the inflow Mach number. For various inflow Mach numbers M_0 the calculations are run until the vortex has passed one third of the flow domain.

In Figure 8.10 the simulation results are shown in form of contour plots of the relative Mach number M/M_0, where the *local* Mach number M is related to the local values of velocity u and v, pressure p and density ρ by

$$M = \sqrt{\frac{u^2 + v^2}{\gamma p/\rho}}.$$

The legends are restricted to minimum and maximum values of M/M_0. In a) the initial condition is shown. The results for the *triangular finite volume cells* turned out to be independent of M_0 and are shown in b) in a single plot. The time steps needed are 811, 7084 and 70000 for the Mach numbers $M_0 = 10^{-1}$, 10^{-2} and 10^{-3}, respectively. The maximum velocity, represented by contour number 10, is reduced by about 50%.

In c) and d) the results for the corner transport upwind method (CTU) on a Cartesian grid with 70×70 cells are shown. The CFL number is about 0.9 and the time steps needed are 268 and 2333 for the Mach numbers $M_0 = 10^{-1}$ and 10^{-2}. The damping of the velocity maximum clearly depends on the Mach number: for $M_0 = 10^{-1}$ the reduction is about 50%, while it is around 80% for $M_0 = 10^{-2}$.

The donor cell upwind method (DCU), results are shown in e) and f), is used on the same Cartesian grid. The CFL numbers was set to 0.45 and time steps were 526 and 6120 for $M_0 = 10^{-1}$ and 10^{-2}. Already in e) the maximum velocity is reduced by about 80% and for $M = 10^{-2}$ the velocity is reduced by more than a factor of 10. The dependency on the Mach number is worse than for the CTU method.

The Gresho vortex simulation shows, that the *accuracy problem* is not limited to steady flow simulations. It occurs on Cartesian grids – for DCU and CTU – and manifests itself in a Mach number dependent dissipation of flow structures. It is absent if the Roe method is used on triangular finite volume cells.

8.5. Inflow of a contact layer

In view of analysing the grid dependency of upwind schemes, we introduce an even simpler test case: the inflow of a contact discontinuity. We investigate two types of steady flow: a steady entropy layer and a steady shear layer, which are set up oblique to the grid lines by the inflow boundary conditions.

The rectangular flow domain $\Omega = [0,1] \times [0,1]$ is in one test series of Cartesian type with 70×70 square cells. For the triangular grid, derived from the Cartesian as depicted in Figure 8.7, a comparable resolution is obtained for $2 \times 50 \times 50$ triangular cells.

8.5.1. Steady entropy layer

All calculations are started with uniform initial conditions

$$\rho_0 = 1.0,$$
$$\mathbf{u}_0 = (u_0, v_0)^T,$$
$$p_0 = 1.0,$$

and are stopped once the residual, given by the sum over all flux differences, indicates steady state. For lower Mach numbers the time to reach steady state increases like $1/M$ since in this test case the steady density distribution is non-uniform due to the inflow boundary conditions.

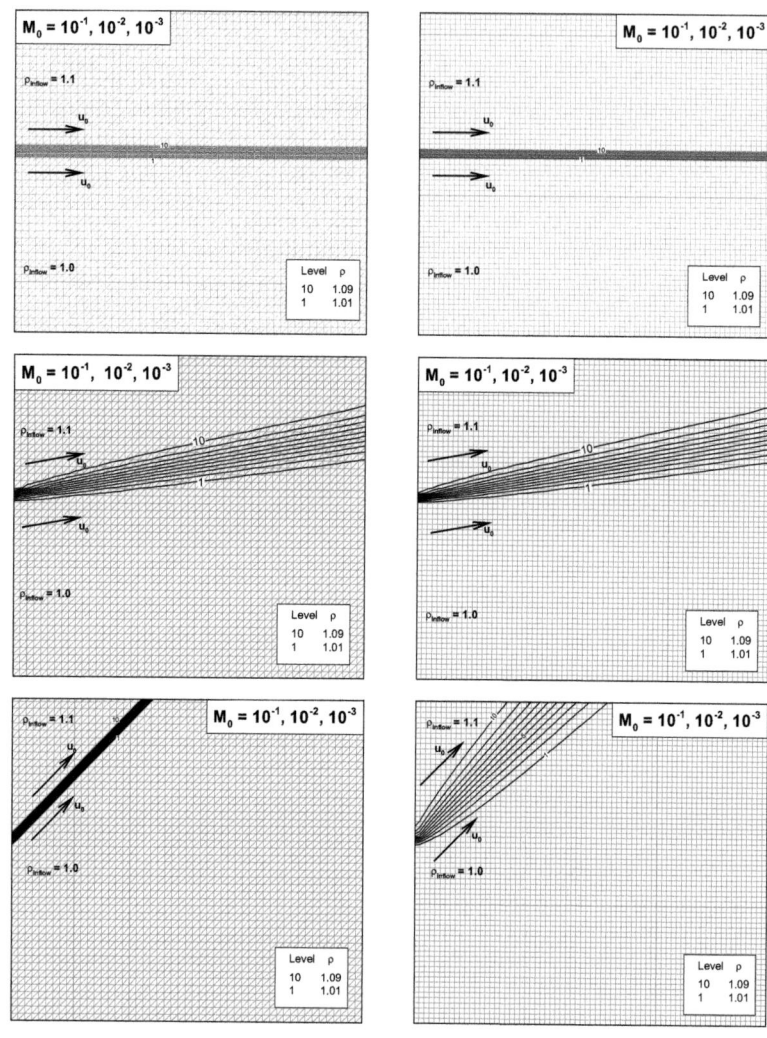

Figure 8.11.: Inflow of an entropy discontinuity on a grid with $2 \times 50 \times 50$ triangular cells (left column) and 70×70 square cells (right column). The inflow angle varies from $\alpha = 0°$ on top, over $\alpha = 10°$ in the middle to $\alpha = 45°$ in the bottom. The results for different Mach numbers are equal and shown in the same diagram.

On the left boundary $\partial\Omega_{\text{left}} = 0 \times [0,1]$ and (only for $\alpha > 0$) on the bottom boundary $\partial\Omega_{\text{bottom}} = [0,1] \times 0$ we impose *inflow boundary conditions* by setting in the ghost cells

$$\rho_{\text{ghost}} = \begin{cases} 1.0 & y \leq 0.5 \\ 1.1 & y > 0.5 \end{cases},$$

$$u_{\text{ghost}} = \|\mathbf{u}_{\text{in}}\| \cos \alpha,$$
$$v_{\text{ghost}} = \|\mathbf{u}_{\text{in}}\| \sin \alpha.$$

The inflow velocity $\mathbf{u}_{\text{in}}$ is equal to the initial velocity $\mathbf{u}_0$ and related to the characteristic Mach number of the test case by:

$$M_0 = \frac{\|\mathbf{u}_0\|}{a_0} = \frac{\|\mathbf{u}_0\|}{\sqrt{\gamma p_0 / \rho_0}} = \frac{\|\mathbf{u}_0\|}{\sqrt{\gamma}}. \tag{8.1}$$

The simulation is run with three different inflow angles

$$\alpha = 0°, \; 10° \text{ and } 45°.$$

Note that for $\alpha = 0°$ the bottom boundary is also treated like an outflow boundary. A simple entropy layer is created by a transported jump in the density: the inflow density in the upper half of the grid is by a factor 1.1 larger than in the lower half. The pressure in the ghost cells is extrapolated from the interior to avoid over-determination.

For the upper and right boundary we use *outflow boundary conditions* by copying the values of the boundary cells into the neighbouring ghost cells. This is a first-order implementation of the von Neumann boundary conditions with zero gradient.

In Figure 8.11 we present the isolines of density for the different inflow angles and grids. The major result of the investigation is that density and entropy distribution are completely independent of the inflow Mach numbers. Therefore the results for the Mach numbers

$$M_0 = 10^{-1}, 10^{-2} \text{ and } 10^{-3}$$

are given in the same diagram. The results on the left of Figure 8.11 are obtained on a grid with $2 \times 50 \times 50 = 5000$ triangular finite volume cells. The grid used for the simulation on the right is a Cartesian grid with $70 \times 70 = 4900$ square cells. It can be observed – for both types of grid – that the smearing of the contact depends on the angle, but there is no dependency on the Mach number. *Therefore, numerical effects in conjunction with entropy transport cannot be the reason for the accuracy problem of the Roe scheme on Cartesian grids in the low Mach number regime.*

8.5.2. Steady shear layer

In Part I we showed that a "cheap" treatment of the shear wave causes incomplete Riemann solvers, such as HLL, to produce large errors in low Mach number computations. In the following test case we investigate the numerical effects of the complete Riemann solver by Roe on a steady, oblique shear layer. The setting for this test case

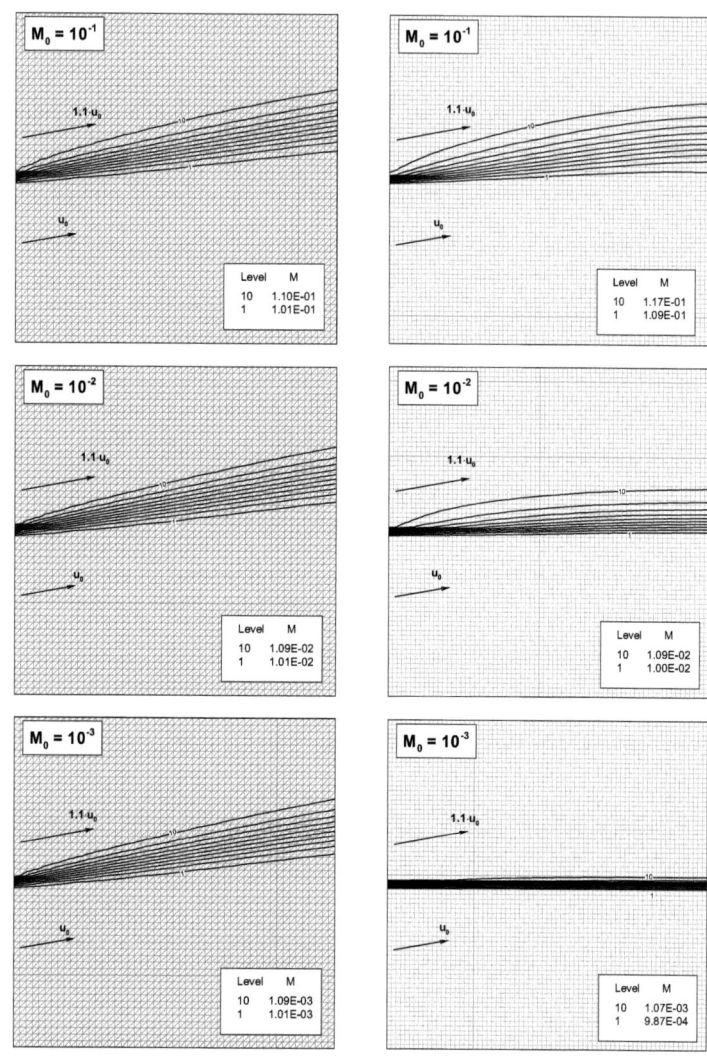

Figure 8.12.: Inflow of a shear discontinuity on a grid with $2 \times 50 \times 50$ triangular cells (left column) and 70×70 square cells (right column). The inflow angle is set to $\alpha = 10°$ and the inflow velocity is set according to the given Mach number ranging from $M_0 = 10^{-1}$ down to $M_0 = 10^{-3}$. In the upper half of the grid the inflow Mach number is 1.1 times larger than in the lower half.

is similar to the entropy layer test case: on the left and bottom edge of the rectangular flow domain $\Omega = [0,1] \times [0,1]$ we choose *inflow boundary conditions* with a jump in the prescribed inflow velocity:

$$\rho_{\text{ghost}} = 1.0\,,$$

$$\mathbf{u}_{\text{ghost}} = \begin{cases} 1.0\,\|\mathbf{u}_{\text{in}}\|(\cos\alpha, \sin\alpha)^T & y \leq 0.5 \\ 1.1\,\|\mathbf{u}_{\text{in}}\|(\cos\alpha, \sin\alpha)^T & y > 0.5 \end{cases},$$

where $\alpha = 10°$ is the angle of inflow in this test case. The absolute value $\|\mathbf{u}_{\text{in}}\|$ is related to the characteristic Mach number M_0 by (8.1). The pressure in the ghost cells p_{ghost} next to the inflow boundary is extrapolated from the neighbouring interior cells.

For the right and upper boundary we impose *outflow boundary conditions* by extrapolating the values of density, velocity and pressure into the ghost cells.

In Figure 8.12, on the left columns, we see the isolines of the Mach number obtained on a grid with $2 \times 50 \times 50$ triangular finite volume cells and on the right column the results for a Cartesian grid with 70×70 cells. In the background the grid lines are indicated in a light grey. We make two major observations:

- *On triangular finite volume cells:* The numerical dissipation is independent of the Mach number and the angle of the shear layer agrees with the inflow boundary condition. Note that numerical experiments with different inflow angles show different grades of dissipation of the contact but the independency of the Mach number remains.

- *On Cartesian cells:* Results obtained with the Roe solver show a bending of the shear layer towards the grid lines for decreasing Mach numbers until it is, for $M_0 \approx 10^{-3}$, completely aligned to the grid.

Obviously, there must be a term in the numerical scheme that treats a pure shear layer with a numerical viscosity of the wrong order of magnitude. This information suggests the way to go for the numerical analysis, which will be presented in the following chapters.

9. Quadrilateral grid cells

The problem of upwind schemes in the low Mach number regime originates in the different orders of magnitude of the transport velocities, which in turn lead to different magnitudes of the numerical damping. In the case of incomplete Riemann solvers such as HLL this causes excessive numerical damping on entropy *and* shear waves – independent of the cell geometry. This facet of the *accuracy problem* can be understood in a one-dimensional setting and was dealt with in Part I of this thesis. To understand the *accuracy problem* of a *complete* Riemann solver such as Roe's, a two-dimensional analysis is necessary. Obviously, this kind of Riemann solver acts locally one-dimensional by decomposing a plane wave – the local Riemann problem – into characteristic components, irrespective of the multi-dimensional nature of the global problem. Imagine an oblique shear layer with constant pressure and density and a divergence-free velocity field. The jump in the velocity field at a cell interface will locally be split into a transverse part defining a shear wave, and a jump in normal direction creating an acoustic wave. The latter is stabilised by the upwind scheme with the absolute value of the sound speed. This introduces the wrong order of magnitude of numerical damping into the transport of the physical shear layer – an effect, which does not cancel across the four edges of a finite volume cell as will be shown. Therefore, the reason for the *accuracy problem* of upwind schemes has to do with the mixing of the characteristic waves, as already proposed by Sesterhenn et al. in [48].

Dimensional considerations exclude a similar effect of over-damping for purely acoustic waves. Pure entropy transport is neither effected, because the mixing of characteristics cannot occur: a jump only in the density variable cannot be part of an acoustic wave, characterised by jumps in pressure and velocity, cf. Sec. 4.1. Therefore we think

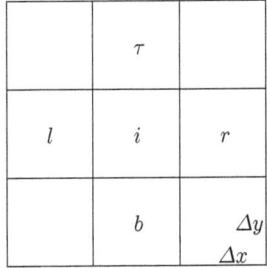

Figure 9.1.: Cartesian grid with cell and edge indices.

it is sufficient to investigate the transport of oblique shear layers to understand the origin of the *accuracy problem*.

We emphasise that the presented analysis in this chapter is not a rigorous proof, but rather a heuristic consideration. The main concern is to uncover the existence of an artificial viscosity of the wrong order of magnitude in a setting, which is as simple as possible.

A rigorous asymptotic analysis for the *accuracy problem* of the Roe scheme on Cartesian grids was given by Viozat in [60]. Therein, the author shows the existence of an unphysical fluctuation of the first-order pressure $p^{(1)}$ under the assumption of a non-trivial velocity field of leading-order $\mathbf{u}^{(0)}$. This result is related to our findings in the sense, that a numerical viscosity of the wrong order induces a pressure field of the wrong order (see Chapter 7 for details).

In Section 9.1 of this chapter we analyse the meaning of a velocity field to be *divergence-free* in a finite-volume sense on a Cartesian grid. The steady transport equation of a divergence-free velocity field of the first-order Roe scheme on a Cartesian grid is discussed in Section 9.2. This analysis will be applied to more general quadrilateral grids in Section 9.3.

9.1. Discrete divergence constraint

We recall a major result of the asymptotic analysis of the Euler equations for low Mach numbers, cf. Equation (2.28):

$$\frac{p_t^{(0)}}{\gamma p^{(0)}} = -\nabla \cdot \mathbf{u}^{(0)} , \qquad (9.1)$$

stating that in a steady flow field the divergence of the leading-order velocity is zero. Therefore we can conclude for a steady low Mach number flow

$$\nabla \cdot \mathbf{u} = \mathcal{O}(\mathrm{M}) \quad \text{as} \quad \mathrm{M} \to 0 .$$

In the context of first-order finite volume methods the velocity is piecewise constant leading to Riemann problems at the cell edges. Note that these Riemann problems are essentially one-dimensional problems and the solution evolves in plane waves. Guillard *et al.* show in [19] that the *exact solution* to a Riemann problem with a jump in the normal velocity component $U^{(0)}$ creates pressure fluctuations in $p^{(1)}$. This agrees with the asymptotic analysis, since in one space dimension divergence and first derivative coincide:

$$\nabla \cdot \mathbf{u}^{(0)} \xrightarrow{\text{1D}} \frac{\partial u^{(0)}}{\partial x}$$

ergo any jump in $U^{(0)}$ at a cell interface creates a first-order pressure $p^{(1)}$. The Roe scheme, as can be expected from any other classical upwind scheme, is shown in [19] to create a pressure of the order $\mathcal{O}(\mathrm{M})$ – which agrees with the asymptotic behaviour of the exact solution. Obviously, this pressure field, created correctly in the one-dimensional setting of the Riemann problem, persists in the global two-dimensional approximation of the flow. We therefore suggest: A necessary condition for the absence of the physically wrong variation in the pressure $p^{(1)}$ is the continuity of the normal component of the leading order velocity $U^{(0)}$ at the cell interfaces. This proposal was exemplarily shown by Guillard in [20] for *Cartesian meshes* and can be summarised as follows:

$$\Delta U^{(0)} \neq 0 \Rightarrow \Delta p^{(1)} \neq 0 \ .$$

In this section we show for the Roe scheme on Cartesian meshes how a jump of the normal component $\Delta U^{(0)}$ leads to an artificial viscosity term of the order $\mathcal{O}(\Delta x/\mathrm{M})$, which in turn induces a non-constant pressure field $p^{(1)}$.

For (special) triangular finite volume cells we prove in Chapter 11

$$\Delta U^{(0)} = 0 \Rightarrow \Delta p^{(1)} = 0 \ ,$$

and the consequences for the degrees of freedom for the leading order velocity $\mathbf{u}^{(0)}$.

9.2. Cartesian grid

In analogy to the numerical experiment in Section 8.5, we can think of a (non-trivial) divergence-free flow as a shear layer which is oblique to the grid. In this setting we assume the pressure p and the density ρ to be constant and identify the Roe averages with the corresponding constants:

$$\hat{p} \rightsquigarrow p \ , \qquad \hat{\rho} \rightsquigarrow \rho \ .$$

Note that the Roe averaged speed of sound $\hat{a}$ is not constant. It deviates from the background sound speed a, defined by $a^2 = \gamma p/\rho$,

$$\hat{a} = a + \mathcal{O}(\mathrm{M}) \ ,$$

as shown in Appendix A.1.

For simplicity we restrict the analysis to a Cartesian grid, where $\Delta x = \Delta y$. A grid detail introducing edge and cell indices is depicted in Figure 9.1. The interface flux of the Roe scheme across edge ri connecting cell i and r is given by

$$\mathbf{f}_{ri}(\hat{\mathbf{q}}_{ri}) = \frac{\mathbf{f}_r + \mathbf{f}_i}{2} - \frac{1}{2}\sum_p \mathbf{r}_p |\lambda_p| \Delta w_p \ ,$$

with $\hat{\mathbf{q}}_{ri}$ denoting the Roe state, $\mathbf{r}_p$ the right eigenvector to the p-characteristic wave, λ_p the corresponding wave speed and Δw_p the jump in the characteristic variable. Details of the Roe scheme can be found in Appendix A.1.

9.2.1. Transport of momentum

We want to know: what is the order of magnitude of the numerical damping in the discrete transport equation of ρu. The analysis can be restricted to the *horizontal* momentum ρu, since the transport equations for horizontal and vertical momentum are equivalent.

The energy equation is not considered here, because the negative effects are expected to become visible for the momentum transport.

Horizontal transport

Using the constancy of p and ρ in the flux of the horizontal momentum $f_{ri}^{\rho u}$ across the edge ri, we obtain

$$\begin{aligned}f_{ri}^{\rho u} &= \frac{(\rho u_r^2 + p) + (\rho u_i^2 + p)}{2} \\ &\quad - \frac{1}{2}(\hat{u}_{ri} - \hat{a}_{ri})|\hat{u}_{ri} - \hat{a}_{ri}|\frac{1}{2\hat{a}_{ri}}(-\rho\Delta_{ri}u) - \frac{1}{2}(\hat{u}_{ri} + \hat{a}_{ri})|\hat{u}_{ri} + \hat{a}_{ri}|\frac{1}{2\hat{a}_{ri}}\rho\Delta_{ri}u \,, \\ &= \overline{\rho u_{ri}^2} + p - \frac{\rho}{2\hat{a}_{ri}}(\bar{u}_{ri}^2 + \hat{a}_{ri}^2)\Delta_{ri}u \,,\end{aligned}$$

(9.2)

where $\hat{u}_{ri}$ denotes the Roe averaged normal velocity and $\bar{u}_{ri}$ its arithmetic mean taken at edge ri. Note that both means coincide for constant density:

$$\hat{u}_{ri} = \frac{\sqrt{\rho_l}u_l + \sqrt{\rho_r}u_r}{\sqrt{\rho_l} + \sqrt{\rho_r}} = \frac{u_l + u_r}{2} = \bar{u}_{ri} \quad \text{for} \quad \rho = \text{const}\,.$$

The term $\Delta_{ri}u$ denotes the jump in u at cell edge ri

$$\Delta_{ri}u := u_r - u_i\,.$$

The mean square of u at the edge ri is defined as

$$\overline{u_{ri}^2} := \frac{u_r^2 + u_i^2}{2}\,.$$

Exploiting the fact that for low Mach numbers $u - a$ is always negative, the flux difference between left and right edge can be written as

$$\begin{aligned}\frac{f_{ri}^{\rho u} - f_{il}^{\rho u}}{\Delta x} &= -\frac{1}{2}\rho\left\{\frac{\bar{u}_{ri}^2\Delta_{ri}u}{\hat{a}_{ri}} - \frac{\bar{u}_{il}^2\Delta_{il}u}{\hat{a}_{il}}\right\}\frac{1}{\Delta x} \\ &\quad + \rho\frac{u_r^2 - u_l^2}{2\Delta x} \\ &\quad - \frac{1}{2}\rho\frac{\hat{a}_{ri}\Delta_{ri}u - \hat{a}_{il}\Delta_{il}u}{\Delta x}\,.\end{aligned}$$

(9.3)

If we scale (9.3) with $\rho_{\text{ref}}u_{\text{ref}}^2$, we notice that the first line on the RHS is a numerical viscosity of order $\mathcal{O}_S(\Delta x \text{M})$, the second line represents the physical transport of order

$\mathcal{O}_S(1)$ and the last line is a numerical viscosity of order $\mathcal{O}_S(\Delta x/M)$ as $M \to 0$. It is this term, which causes the upwind scheme to lose accuracy for decreasing Mach numbers. We interpret this order relation in the context of the entire modified equation for the transport of ρu.

Vertical transport of ρu

For the modified equation, the flux terms for the *vertical* transport of ρu are needed, too. The problem is equivalent to the horizontal transport of momentum ρv given by

$$f_{ri}^{\rho v} = f_{\text{central}}^{\rho v} - \tfrac{1}{2}\left\{\hat{v}_{ri}|\hat{u}_{ri} - \hat{a}_{ri}|\left(-\frac{\rho}{2\hat{a}_{ri}}\Delta_{ri}u\right) + |\hat{u}_{ri}|\rho\Delta_{ri}v + \hat{v}_{ri}|\hat{u}_{ri} + \hat{a}_{ri}|\left(\frac{\rho}{2\hat{a}_{ri}}\Delta_{ri}u\right)\right\}.$$

With the fact that $\hat{u} - \hat{a}$ is negative and $\hat{u} + \hat{a}$ is positive for $M < 1$, the flux can be simplified to

$$f_{ri}^{\rho v} = f_{\text{central}}^{\rho v} - \frac{1}{2}\left\{\frac{\rho}{\hat{a}_{ri}}\hat{u}_{ri}\hat{v}_{ri}\Delta_{ri}u + \rho|\hat{u}_{ri}|\Delta_{ri}v\right\}.$$

The transport of ρu in vertical direction at cell edge τi is obtained from this equation by identifying u with v:

$$f_{\tau i}^{\rho u} = \rho(\overline{uv})_{\tau i} - \frac{1}{2}\left\{\frac{\rho}{\hat{a}_{\tau i}}\hat{v}_{\tau i}\hat{u}_{\tau i}\Delta_{\tau i}v + \rho|\hat{v}_{\tau i}|\Delta_{\tau i}u\right\}, \quad (9.4)$$

where the arithmetic mean $(\overline{uv})_{\tau i}$ of the velocity product uv for the values separated by edge τi is defined as

$$(\overline{uv})_{\tau i} := \frac{(uv)_\tau + (uv)_i}{2}.$$

For the difference of the interface fluxes at top edge τi and bottom edge bi, we find

$$\begin{aligned}\frac{f_{\tau i}^{\rho u} - f_{ib}^{\rho u}}{\Delta y} &= -\frac{1}{2}\rho\left\{\frac{\bar{u}_{\tau i}\bar{v}_{\tau i}\Delta_{\tau i}v}{\hat{a}_{\tau i}} - \frac{\bar{u}_{ib}\bar{v}_{ib}\Delta_{ib}v}{\hat{a}_{ib}}\right\}\frac{1}{\Delta y} \\ &\quad + \rho\frac{u_\tau v_\tau - u_b v_b}{2\Delta y} \\ &\quad - \frac{1}{2}\rho\frac{|\bar{v}_{\tau i}|\Delta_{\tau i}u - |\bar{v}_{ib}|\Delta_{ib}u}{\Delta y}.\end{aligned} \quad (9.5)$$

The absence of a factor $\rho\hat{a}$ indicates that there is no numerical viscosity of order $\mathcal{O}_S(\Delta x/M)$. We exploit this fact in the analyses for other cell geometries by focusing only on the transport of momentum components normal to an edge.

Interpretation with the modified equation

The modified equation approach can give some insight in the numerical behaviour of the scheme. To this end we put the intermediate results (9.3) and (9.5) into a single equation and introduce the standard scaling by $\rho_{\text{ref}}u_{\text{ref}}^2$. Assuming a smooth solution to the original PDE, we obtain the following expression for the (steady) modified equation:

$$\rho(u^2)_x + \rho(uv)_y = M \left\{ \frac{1}{2}\rho\frac{u^2}{a}u_{xx}\Delta x + \rho\frac{u}{a}u_x^2\Delta x - \frac{1}{2}\frac{\rho}{a}\{uvv_{yy} + (uv)_y v_y\}\Delta y \right\}$$
$$+ \left\{ \frac{1}{2}\rho v u_{yy}\Delta y + \rho v_y u_y \Delta y \right\} \quad (9.6)$$
$$+ \frac{1}{M}\left\{ \frac{1}{2}\rho a u_{xx}\Delta x \right\}.$$

The viscous terms fall into three orders of magnitude: $\mathcal{O}_S(M\Delta x)$, $\mathcal{O}_S(\Delta x)$ and $\mathcal{O}_S(\Delta x/M)$. This relation can be summarised in the order relation for the numerical Reynolds number of the transport of momentum ρu in x-direction:

$$\boxed{\text{Re}^{\text{Roe}}_{\rho u,x} = \mathcal{O}_S\left(\frac{M}{\Delta x}\right)}.$$

The transport of momentum ρu in y-direction is damped with the correct order of magnitude:

$$\boxed{\text{Re}^{\text{Roe}}_{\rho u,y} = \mathcal{O}_S\left(\frac{1}{\Delta y}\right)}.$$

Conclusion

We conclude: the presence of a second order difference in u of order $\mathcal{O}_S(1)$ leads to an artificial viscosity of the first-order Roe scheme of the order $\mathcal{O}_S(\Delta x/M)$ as $M \to 0$. Even in the absence of acoustic waves does the Roe scheme introduce artificial viscosity of the order corresponding to these waves. Thus, shear layers, which are oblique to the grid, cannot be calculated accurately for decreasing Mach numbers. This comprises all practically relevant flows. One exception is the potential stagnation point flow

$$u = ax,$$
$$v = -ay$$

with $u_{xx} = 0$ and $v_{yy} = 0$, so that in the stagnation point region of a compressible flow around an aerofoil the deviation from the true solution is probably not too severe.

In the beginning of this chapter we showed that the discrete divergence constraint is only satisfied, if there are no jumps of the normal component of the velocity at a cell interface. If a velocity field is divergence-free in this sense, the Roe scheme has no artificial viscosity terms of the order $\mathcal{O}_S(\Delta x/M)$ as $M \to 0$. But for this trivial case there in no need for calculations any way.

9.3. Generalisation

In this section we generalise the results obtained for Cartesian grids. In the hierarchy of generalisations there are rectilinear grids with parallelogram cells as depicted in

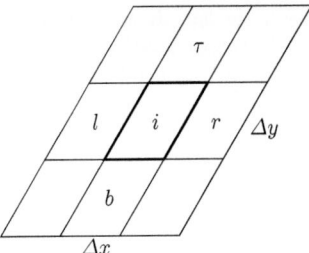

Figure 9.2.: Grid with parallelogram cells and corresponding indices.

Figure 9.2 and, more general, cells with only two parallel edges, trapezoidal cells and then arbitrary quadrilaterals.

We restrict the analysis to rectilinear grids here, cf. Figure 9.2, but the results suggest that the *accuracy problem* is caused by the presence of parallel cell edges.

Since the *accuracy problem* occurs if artificial viscosity of the order $\mathcal{O}_S(\Delta x/M)$ acts on any component of the momentum, it will suffice to analyse:

1. the transport of horizontal momentum ρu and

2. the numerical flux terms of this order, called f $[\mathcal{O}_S(1/M)]$.

We make use of the following notation and refer to Figure 9.2 for details concerning edge and cell indices:

$\mathbf{n}^{il} = (n_1^{il}, n_y^{il})^T$ is the unit normal vector of edge il pointing from cell i to l.

$\mathbf{t}^{il} = (-n_y^{il}, n_x^{il})^T$ is the corresponding unit transverse vector.

$\Delta_{il} U = U_i - U_l$ is the jump of the normal component of the velocity at il.

The velocity $\mathbf{u}$ can be expanded in the local coordinate system $(\mathbf{n}, \mathbf{t})$ and the global coordinate system $(\mathbf{e}_1, \mathbf{e}_2)$:

$$\mathbf{u} = U\mathbf{n} + V\mathbf{t} = u\mathbf{e}_1 + v\mathbf{e}_2 = (u, v)^T.$$

The equation of momentum update in semi-discrete form with the physical convective flux difference ΔF^{conv} and the flux of numerical viscosity ΔF^{visc} can be written as

$$\frac{\partial}{\partial t}\rho u + \frac{1}{A_\square}\Delta F^{\text{conv}} = \frac{1}{A_\square}\Delta F^{\text{visc}}, \tag{9.7}$$

with the surface area of the finite volume cell $A_\square$. We want to find the part of ΔF^{visc} which is $\mathcal{O}_S(1/M)$. The fluxes of the momentum update of ρu only contain $\mathcal{O}_S(1/M)$ terms across the left edge il and right edge ri, because the transverse transport of

ρu represents a pure shear wave in the local Riemann problems at ri and bi. Of the velocity jump $\Delta \mathbf{u}$ only the normal component $\Delta U = \mathbf{n} \cdot \Delta \mathbf{u}$ contributes with $\mathcal{O}_S(1/M)$ to the horizontal momentum. Since the *horizontal* flux component is considered, only the fraction n_x enters the equation. With theses generalisations of equation (9.2), we obtain

$$f_{il}^{\rho u}[\mathcal{O}_S(1/M)] = -\frac{1}{2}\rho a n_x^{il} \mathbf{n}^{il}(\mathbf{u}_l - \mathbf{u}_i) = +\frac{1}{2}\rho a n_x^{il}\Delta_{il}U ,$$

$$f_{ri}^{\rho u}[\mathcal{O}_S(1/M)] = -\frac{1}{2}\rho a n_x^{ir} \mathbf{n}^{ir}(\mathbf{u}_r - \mathbf{u}_i) = -\frac{1}{2}\rho a n_x^{ir}\Delta_{ri}U .$$

The $\mathcal{O}_S(1/M)$ part of the flux difference $\Delta F^{\text{visc}} = (f_{ri}^{\rho u} - f_{il}^{\rho u})\Delta y$ is given by

$$\Delta F^{\text{visc}}[\mathcal{O}_S(1/M)] = \frac{1}{2}\rho a \left\{ n_x^{ir}\Delta y \mathbf{n}^{ir} \cdot (\mathbf{u}_r - \mathbf{u}_i) - n_x^{il}\Delta y \mathbf{n}^{il} \cdot (\mathbf{u}_i - \mathbf{u}_l) \right\} .$$

This can be simplified, using the fact that the left and right edge are parallel, $\mathbf{n}^{il} = -\mathbf{n}^{ir}$, to the equation

$$\boxed{\Delta F^{\text{visc}}[\mathcal{O}_S(1/M)] = \frac{1}{2}\rho a \Delta y n_x^{ir} \mathbf{n}^{ir} \cdot (\mathbf{u}_l - 2\mathbf{u}_i + \mathbf{u}_r) .} \qquad (9.8)$$

The flux contribution to the update equation is obtained by dividing (9.8) by the cell area $A_\square = \Delta x \Delta y \cos(\mathbf{n}^{ib}, \mathbf{n}^{ir})$.

Equation (9.8) contains a second order difference of the flow velocity. It only vanishes if the velocity components perpendicular to edge ri have equal (discrete) gradients, which implies some sort of alignment to the grid. As long as the second order differences do not vanish and are $\mathcal{O}_S(1)$ as $M \to 0$, the numerical dissipation of the momentum ρu outweighs the physical transport. This can be expressed by the numerical Reynolds numbers for the transport of momentum ρu along the x- and y-axis:

$$\boxed{\text{Re}_{\rho u,x}^{\text{Roe}} = \mathcal{O}_S(\tfrac{M}{\Delta x}), \quad \text{Re}_{\rho u,y}^{\text{Roe}} = \mathcal{O}_S(\tfrac{1}{\Delta y}).} \qquad (9.9)$$

In this case an unphysical pressure field of order $\mathcal{O}_S(M)$ is induced by the scheme, as was shown in Chapter 7. For the transport of ρu along the y-axis we made use of the fact that ρu has no normal component to edge τi or bi, and consequently, no $\mathcal{O}_S(1/M)$-contribution to the artificial viscosity.

Conclusion

If a grid cell has a pair of parallel interfaces, the corresponding normal component of the velocity U experiences a damping of the order $\mathcal{O}(1/M)$ as $M \to 0$. In a steady solution, this effect can disappear if the second order difference of U vanishes. This happens in either of two cases:

1. Either this constraint on U is enforced by the scheme against the physics of the flow determined by the boundary conditions. This enforcement is balanced by an unphysical pressure field of order $\mathcal{O}(M)$ as $M \to 0$.

2. Or this constraint is in agreement with the physics – but then the physically correct flow is trivial or aligned to the grid.

10. Triangular grid cells

In Section 10.1 we apply the analysis, presented in the previous chapter, to the momentum transport by the Roe scheme on *triangular* finite volume cells. For simplicity we consider special cells, which originate from squares by introducing a diagonal, as depicted in Figure 10.1.

Section 10.2 is dedicated to dual grids of triangulations, which have *hexagonal* finite volume cells.

10.1. Triangulation derived from a Cartesian grid

Any physical transport phenomenon will experience two different kinds of triangular cells on its way: an "upper" triangle such as cell i and a "lower" triangle such as cell d, l or τ; see Figure 10.1 for details. To understand the numerical effects on the momentum transport it is useful to derive equations for both sorts of triangular cells and to compare them.

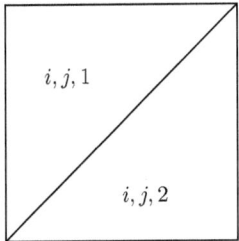
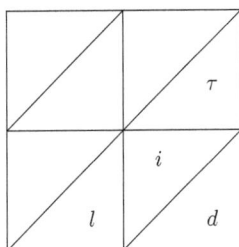

Figure 10.1.: Triangular cells originating from Cartesian grid cells by introducing a diagonal. The global indices (left) are used for the addressing in the numerical algorithm. The local indices (right) are used for the analysis.

10.1.1. Upper triangle

For the flux across the diagonal edge di we make use of the following relations for the velocity $\mathbf{u} = (u, v)^T$, valid for the special cell geometry used in this section:

$$U = \frac{1}{\sqrt{2}}(u - v), \quad V = \frac{1}{\sqrt{2}}(u + v), \tag{10.1}$$

$$\Rightarrow \hat{U} = \bar{U} = \frac{1}{\sqrt{2}}(\bar{u} - \bar{v}) \quad \text{and} \quad \hat{V} = \bar{V} = \frac{1}{\sqrt{2}}(\bar{u} + \bar{v}), \tag{10.2}$$

where U and V are the normal and transverse velocity components. Recall that Roe average $\hat{u}$ and arithmetic mean $\bar{u}$ coincide for constant density. For a constant variable the Roe average is also a constant. For the pressure p and the density ρ, which are assumed constant, we therefore identify

$$\hat{p} \rightsquigarrow p, \quad \hat{\rho} \rightsquigarrow \rho.$$

For index notation see also Figure 10.2. The flux $f_{di}^{\mathbf{n}}$ of normal momentum ρU across the diagonal di is given by

$$f_{di}^{\mathbf{n}} = \frac{1}{2}(f_l^{\mathbf{n}} + f_r^{\mathbf{n}}) - \frac{1}{2}\sum_p \mathbf{r}_p^{(2)} |\lambda_p| \Delta w_p$$

$$= \frac{1}{2}\{\rho(U_l)^2 + \rho(U_r)^2\} + p$$

$$- \frac{1}{2}\{(\hat{U}_{di} - \hat{a}_{di})|\hat{U}_{di} - \hat{a}_{di}|\left(-\frac{1}{2\hat{a}_{di}}\rho\Delta_{di}U\right) + (\hat{U}_{di} + \hat{a}_{di})|\hat{U}_{di} + \hat{a}_{di}|\left(\frac{1}{2\hat{a}_{di}}\rho\Delta_{di}U\right)\}$$

$$= \frac{1}{2}\rho\{(U_l)^2 + (U_r)^2\} + p - \frac{1}{2}\frac{(\hat{U}_{di})^2 + \hat{a}_{di}^2}{\hat{a}_{di}}\rho\Delta_{di}U,$$

where we have used $\hat{a} > \hat{U}$ for low Mach numbers to eliminate the absolute values of $|\hat{U} - \hat{a}|$ and $|\hat{U} + \hat{a}|$. The term $\Delta_{di}U$ denotes the jump of the normal velocity at edge di. For the flux $f_{di}^{\mathbf{t}}$ of the transverse momentum ρV across the diagonal edge di we obtain

$$f_{di}^{\mathbf{t}} = \frac{1}{2}(f_l^{\mathbf{t}} + f_r^{\mathbf{t}}) - \frac{1}{2}\sum_p \mathbf{r}_p^{(3)} |\lambda_p| \Delta w_p$$

$$= \frac{\rho U_l V_l + \rho U_r V_r}{2} - \frac{1}{2}\{\hat{V}_{di}|\hat{U}_{di} - \hat{a}_{di}|\left(-\frac{\rho\Delta_{di}U}{2\hat{a}_{di}}\right) + |\hat{U}_{di}|\rho\Delta_{di}V + \hat{V}_{di}|\hat{U}_{di} + \hat{a}_{di}|\frac{\rho\Delta_{di}U}{2\hat{a}_{di}}\}$$

$$= \rho\frac{U_l V_l + U_r V_r}{2} - \frac{1}{2}\{\frac{\hat{V}_{di}\hat{U}_{di}}{\hat{a}_{di}}\rho\Delta_{di}U + |\hat{u}_{di}^{\mathbf{n}}|\rho\Delta_{di}V\}.$$

The flux components of the local coordinate system can be combined similar to (10.1) to obtain an expression for the horizontal momentum flux $f^{\rho u}$:

$$f_{di}^{\rho u} = \frac{1}{\sqrt{2}}(f_{di}^{\mathbf{n}} + f_{di}^{\mathbf{t}}),$$

$$= \frac{1}{\sqrt{2}}\{\rho\overline{(U)^2_{di}} + p + \rho\frac{U_l V_l + U_r V_r}{2} - \frac{1}{2}\rho\frac{\overline{(U)^2_{di}} + \hat{a}_{di}^2 + \bar{V}_{di}\bar{U}_{di}}{\hat{a}_{di}}\Delta U - \frac{1}{2}\rho|\bar{U}_{di}|\Delta_{di}V\}.$$

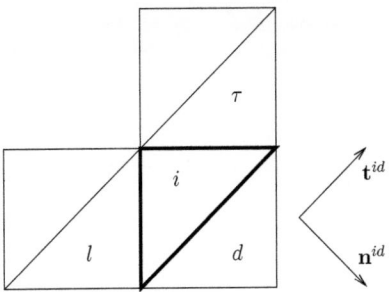

Figure 10.2.: Detail of a triangulation based on a Cartesian grid.
Left: edge and cell indices for the analysis of an upper triangle.
Right: local coordinate system for the diagonal edge di.

If we further express the local velocity coordinates U, V by the globally defined u and v, we obtain:

$$f_{di}^{\rho u} = \frac{1}{\sqrt{2}} \left\{ \underbrace{\frac{\rho}{2}[\frac{(u_l-v_l)^2}{2}+\frac{(u_r-v_r)^2}{2}+\frac{u_l^2-v_l^2}{2}+\frac{u_r^2-v_r^2}{2}]}_{T}+p \right. \tag{10.3}$$
$$\left. -\underbrace{\frac{\rho}{2\hat{a}_{di}}[\frac{(\bar{u}_{di}-\bar{v}_{di})^2}{2}+\hat{a}_{di}^2+\frac{\bar{u}_{di}^2-\bar{v}_{di}^2}{2}]}_{V}\Delta_{di}U - \frac{\rho}{2}\frac{|\bar{u}_{di}-\bar{v}_{di}|}{\sqrt{2}}\Delta_{di}V \right\}.$$

The transport term T can be simplified as follows

$$T = \frac{\rho}{4}[u_l^2 - 2u_l v_l + v_l^2 + u_r^2 - 2u_r v_r + v_r^2 + u_l^2 - v_l^2 + u_r^2 - v_r^2]$$
$$= \frac{\rho}{4}[2u_l^2 - 2u_l v_l + 2u_r^2 - 2u_r v_r] = \frac{\rho}{2}[u_l(u_l - v_l) + u_r(u_r - v_r)]$$
$$= \rho[(\overline{u^2})_{di} - (\overline{uv})_{di}].$$

For the viscosity term V we obtain

$$V = \frac{\rho}{4\hat{a}_{di}}[2\hat{a}_{di}^2 + \bar{u}_{di}^2 - 2\bar{u}_{di}\bar{v}_{di} + \bar{v}_{di}^2 + \bar{u}_{di}^2 - \bar{v}_{di}^2] = \frac{\rho}{4\hat{a}_{di}}[2\hat{a}_{di}^2 + 2\bar{u}_{di}^2 - 2\bar{u}_{di}\bar{v}_{di}]$$
$$= \frac{\rho}{2\hat{a}_{di}}[\hat{a}_{di}^2 + \bar{u}_{di}(\bar{u}_{di} - \bar{v}_{di})].$$

Inserting these simplified terms for T and V into (10.3) gives

$$\boxed{f_{di}^{\rho u} = \frac{1}{\sqrt{2}}\left\{\rho[\overline{u^2}_{di} - (\overline{uv})_{di}] + p - \frac{1}{2}\frac{\rho}{\hat{a}_{di}}[\hat{a}_{di}^2 + \bar{u}_{di}^2 - \bar{u}_{di}\bar{v}_{di}]\Delta_{di}U - \frac{1}{2}\rho|\bar{u}_{di} - \bar{v}_{di}|\frac{\Delta_{di}V}{\sqrt{2}}\right\}} \tag{10.4}$$

The equations for the flux across edge li (9.2) and τi (9.4) were derived in Section 9.2 and are stated again for completeness:

$$f_{il}^{\rho u} = \rho(\overline{u^2})_{il} + p - \frac{1}{2}\frac{\rho}{\hat{a}_{il}}(\bar{u}_{il}^2 + \hat{a}_{il}^2)\Delta_{il}u \ , \qquad (10.5)$$

$$f_{\tau i}^{\rho u} = \rho(\overline{uv})_{\tau i} - \frac{1}{2}\rho|\bar{v}_{\tau i}|\Delta_{\tau i}u - \frac{1}{2}\frac{\rho}{\hat{a}_{\tau i}}\bar{u}_{\tau i}\bar{v}_{\tau i}\Delta_{\tau i}v \ . \qquad (10.6)$$

The velocity jump at the left edge li, with the normal vector pointing in the opposite direction to the x-axis, has to be calculated as follows:

$$\Delta u|_{\text{left edge}} = (-u_l) - (-u_i) = u_i - u_l = \Delta_{il}u \ .$$

Using all three fluxes from the triangular cell, the semi-discrete equation for the transport of horizontal momentum ρu reads

$$\frac{\partial}{\partial t}\rho u + \frac{1}{A_\triangle}\left\{f_{\tau i}^{\rho u}\Delta x - f_{il}^{\rho u}\Delta y + f_{di}^{\rho u}\Delta z\right\} = 0 \ . \qquad (10.7)$$

The flux terms herein can again be split into a convective and viscous part

$$\frac{\partial}{\partial t}\rho u + \frac{1}{A_\triangle}\Delta F^{\text{conv}} = \frac{1}{A_\triangle}\Delta F^{\text{visc}} \ , \qquad (10.8)$$

where $A_\triangle = \frac{1}{2}\Delta x \Delta y$ is the cell surface area.

Convective flux difference The convective flux difference can be written as

$$\Delta F^{\text{conv}} = \{\rho(\overline{uv})_{\tau i} - \rho\overline{u^2}_{il} - p + \rho[\overline{u^2}_{di} - (\overline{uv})_{di}] + p\}\Delta x \ , \qquad (10.9)$$

$$\Rightarrow \frac{\Delta F^{\text{conv}}}{A_\triangle} = \rho\frac{(\overline{uv})_{\tau i} - (\overline{uv})_{di}}{\frac{1}{2}\Delta x} + \rho\frac{\overline{u^2}_{di} - \overline{u^2}_{il}}{\frac{1}{2}\Delta x} \ . \qquad (10.10)$$

We assume the existence of smooth solution u and v to the original PDE, which is sensible for low Mach numbers. We can express the discrete expressions in terms of derivatives using Taylor expansions about the barycentre x_i of cell i:

$$(\overline{uv})_{\tau i} = \frac{1}{2}(u_\tau v_\tau + u_i v_i) \ ,$$

$$u_\tau = u_i + (\frac{1}{3}u_x\Delta x + \frac{2}{3}u_y\Delta y)|_{x_i} + \ldots \ ,$$

$$v_\tau = v_i + (\frac{1}{3}v_x\Delta x + \frac{2}{3}v_y\Delta y)|_{x_i} + \ldots \ ,$$

$$\Rightarrow u_\tau v_\tau = (uv)_i + (\frac{1}{3}uv_x + \frac{2}{3}uv_y + \frac{1}{3}vu_x + \frac{2}{3}vu_y)|_{x_i}\Delta x + \ldots \ ,$$

$$\Rightarrow (\overline{uv})_{\tau i} = (uv)_i + (\frac{1}{6}uv_x + \frac{1}{3}uv_y + \frac{1}{6}vu_x + \frac{1}{3}vu_y)|_{x_i}\Delta x + \ldots \ ,$$

where we made use of the fact that $\Delta x = \Delta y$ in a Cartesian grid. For the term $(\overline{uv})_{di}$ we obtain in a similar way

$$\overline{uv}_{di} = \frac{1}{2}(u_d v_d + u_i v_i),$$

$$u_d = u_i + (\frac{1}{3}u_x \Delta x - \frac{1}{3}u_y)|_{x_i}\Delta y + \cdots,$$

$$v_d = v_i + (\frac{1}{3}v_x \Delta x - \frac{1}{3}v_y \Delta y)|_{x_i} + \cdots,$$

$$\Rightarrow u_d v_d = (uv)_i + (\frac{1}{3}uv_x - \frac{1}{3}uv_y + \frac{1}{3}vu_x - \frac{1}{3}vu_y)|_{x_i}\Delta x + \cdots,$$

$$\Rightarrow \overline{uv}_{di} = (uv)_i + (\frac{1}{6}uv_x - \frac{1}{6}uv_y + \frac{1}{6}vu_x - \frac{1}{6}vu_y)|_{x_i}\Delta x + \cdots,$$

where, again, we used the fact $\Delta x = \Delta y$ for Cartesian grids in the last two implications. The second term on the RHS of equation (10.10) can be expanded in an analogous way. Inserting all Taylor expansions into (10.10), we obtain the following modified expression for the convective flux

$$\frac{\Delta F^{\text{conv}}}{A_\Delta} = \rho(u^2)_x + \rho(uv)_y + \mathcal{O}(\Delta x) \quad \text{as} \quad \Delta x \to 0.$$

This is indeed the physical convection for the transport of horizontal momentum ρu in the absence of pressure gradients plus a first-order truncation error.

Viscous flux difference Using equations (10.4) to (10.6) for the transport of horizontal momentum ρu across the edges we obtain for the artificial viscosity flux:

$$\Delta F_i^{\text{visc}} = \frac{1}{2}\rho \left\{ \frac{\bar{u}_{\tau i}\bar{v}_{\tau i}\Delta_{\tau i}v}{\hat{a}_{\tau i}} - \frac{\bar{u}_{il}^2 \Delta_{il}u}{\hat{a}_{il}} + \frac{1}{\hat{a}_{di}}(\bar{u}_{di}^2 - \bar{u}_{di}\bar{v}_{di})\frac{\Delta_{di}u - \Delta_{di}v}{\sqrt{2}} \right\}\Delta x$$
$$+ \frac{1}{2}\rho \left\{ |\bar{v}_{\tau i}|\Delta_{\tau i}u + |\bar{u}_{di} - \bar{v}_{di}|\frac{\Delta_{di}u + \Delta_{di}v}{2} \right\}\Delta x \qquad (10.11)$$
$$+ \frac{1}{2}\rho \left\{ -\hat{a}_{il}\Delta_{il}u + \hat{a}_{di}\Delta_{di}U \right\}\Delta x.$$

If we scale the transport equation (10.8) in a way that the physical transport is of order one, i.e. dividing it by $\rho_{\text{ref}} u_{\text{ref}}^2$, the terms on the RHS of (10.11) are $\mathcal{O}_S(M)$, $\mathcal{O}_S(1)$ and $\mathcal{O}_S(1/M)$ on consecutive lines. Only the viscosity terms of order $\mathcal{O}_S(1/M)$, denoted by $\Delta F^{\text{visc}}[\mathcal{O}_S(1/M)]$ in the following, can affect the numerical results for decreasing Mach numbers. We summarise the result in the equation

$$\boxed{\Delta F_i^{\text{visc}}[\mathcal{O}_S(1/M)] = \frac{1}{2}\rho \left\{ \hat{a}_{di}\Delta_{di}U - \hat{a}_{il}\Delta_{il}u \right\}\Delta x} \qquad (10.12)$$

which states: *the momentum transport in a flow with constant pressure and density experiences a numerical viscosity of $\mathcal{O}_S(1/M)$, if there are jumps in the normal component of the velocity field at the cell edges, which do not cancel each other.*

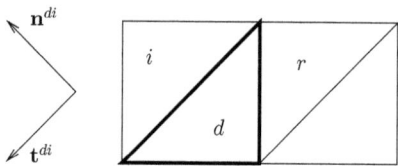

Figure 10.3.: Right: cell indices for the analysis of the lower triangle.
Left: local coordinate system for the diagonal edge di.

10.1.2. Lower triangle

Owing to the regularity of the grid, all calculated fluxes of the previous section can be re-used for the analysis of the lower triangle d, albeit with opposite signs and indices as depicted in Figure 10.3. The semi-discrete equation

$$\frac{\partial}{\partial t}\rho u + \frac{1}{A_\triangle}(f_{rd}\Delta y - f_{db}\Delta x - f_{di}\Delta z) = 0 ,$$

with index db representing the lower interface of cell d, is split into convective and viscous parts

$$\frac{\partial}{\partial t}\rho u + \frac{\Delta F^{\mathrm{conv}}}{A_\triangle} = \frac{\Delta F^{\mathrm{visc}}}{A_\triangle} .$$

Again only the viscous flux term of order $\mathcal{O}_S(1/M)$, given by

$$\boxed{\Delta F_d^{\mathrm{visc}}\left[\mathcal{O}_S(1/M)\right] = \frac{1}{2}\rho(\hat{a}_{ri}\Delta_{rd}u - \hat{a}_{di}\Delta_{di}U)\Delta x} \qquad (10.13)$$

is of interest. The differences

$$\Delta_{rd}u = u_r - u_d ,$$
$$\Delta_{di}U = U_d - U_i ,$$

are again the jumps in the normal component of the velocity at the corresponding cell edges.

Remark

It is easy to see from the previous derivation that for general triangular finite volume cells the flux of artificial viscosity of order $\mathcal{O}_S(\Delta x/M)$ contains all three jumps of the normal component U at the three cell interfaces. Before showing in the next chapter that these jumps actually vanish (with a physically correct flow field), we take a closer look at the implications of a divergence-free, piecewise constant velocity field in the finite volume sense.

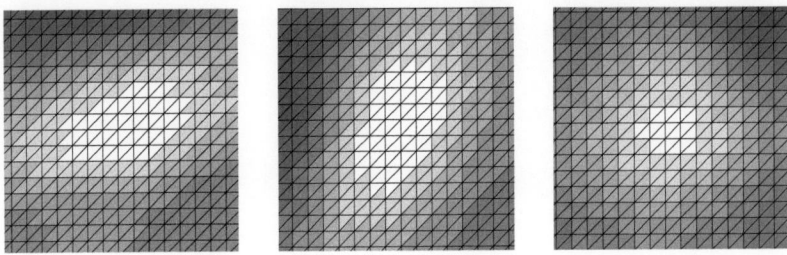

Figure 10.4.: Detail of a Gresho vortex simulation at $M_0 = 10^{-3}$ with the first-order Roe scheme. The grey-scale contour plots show individual components of the piecewise constant velocity field (not interpolated raw data). Left: horizontal components u. Middle: vertical component v. Right: component normal to the diagonal U.

10.1.3. Discrete divergence constraint

In the previous chapter we suggested that the Roe scheme only avoids a physically wrong pressure field $p^{(1)}$, if the discrete (one-dimensional) divergence constraint

$$\Delta U^{(0)} = 0$$

is satisfied. Before we prove this fact in the following chapter, we give some numerical evidence that the scheme actually creates such a velocity field with continuous normal components at the cell interfaces. Figure 10.4 shows contours of velocity components from a Gresho vortex simulation with initial Mach number $M_0 = 10^{-3}$. The pairing of the horizontal component u (left), the vertical component v (middle) as well as the pairing of the component normal to the diagonal U (right) is clearly visible.

A study of this effect with initial Mach numbers ranging from $M_0 = 10^{-1}$ down to $M_0 = 10^{-6}$ is depicted in the diagrams of Figure 10.5. For a detailed description of the numerical experiment see Section 8.4. Besides the jump in the normal components of the velocity, the vorticity, calculated with central differences

$$\omega = \frac{v_d - v_l}{\Delta x} - \frac{u_\tau - u_d}{\Delta y},$$

and the kinetic energy density

$$e_{\text{kin}} = \frac{1}{2}\rho(u_i^2 + v_i^2)$$

is plotted. The diagram shows the average of these quantities across a central region to exclude effects from the boundary. All calculations were stopped after 1000 time steps, which represents a fixed time on the acoustic time scale. Note that in the simulation software the speed of sound is fixed through background values $p_0 = 1$, $\rho_0 = 1$ and the

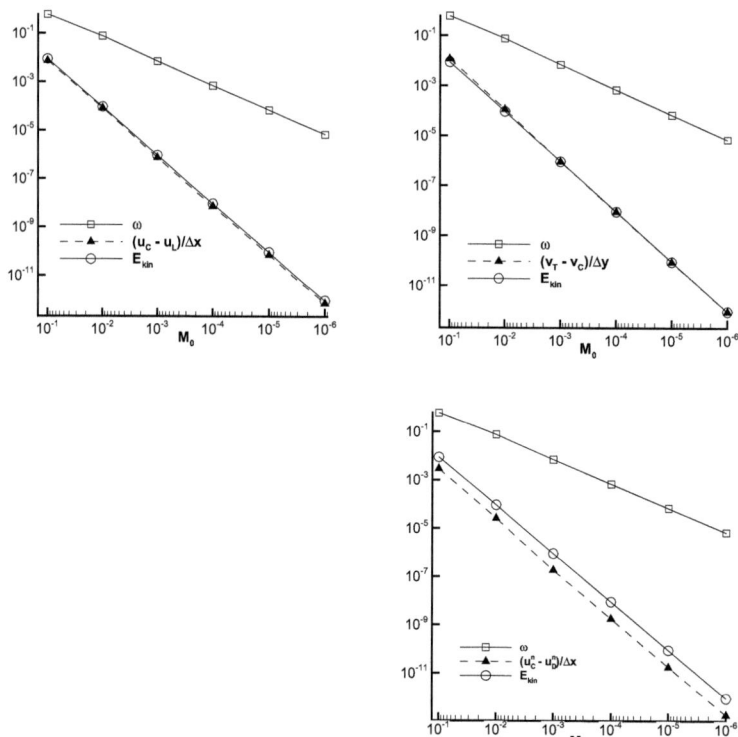

Figure 10.5.: Gresho vortex simulation with the first-order Roe scheme for various Mach numbers. Comparison between the average jump of the velocity component normal to a cell edge, the vorticity ω and the kinetic energy e_{kin} after 1000 time steps. The average jump in u (top), v (middle) and the velocity component normal to the square diagonal U (bottom) are all $\mathcal{O}_S(M^2)$.

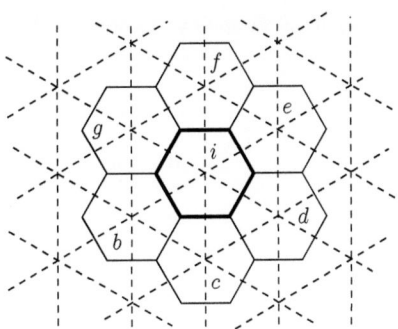

Figure 10.6.: Triangulation (dashed line) with corresponding dual grid of regular hexagons (solid line) with corresponding cell indices.

flow velocity is adapted to meet the Mach number M_0. As a consequence $u = \mathcal{O}_S(M_0)$ and the same is true for any central derivatives thereof, such as the vorticity w, as can be seen in Diagram 10.5. The kinetic energy, containing the square u^2, is $\mathcal{O}_S(M^2)$. Obviously, the same order relation holds for the jump in the normal components at the vertical edge: $u_i - u_l$ (top), the horizontal edge: $v_\tau - v_i$ (middle) and the diagonal edge $U_i - U_d$ (bottom). Thus, the velocity field is divergence-free in the discrete, one-dimensional sense to an order of $\mathcal{O}_S(M^2)$ or, with the standard scaling used in the presented analysis, to an order of $\mathcal{O}_S(M)$.

10.2. Dual grids of triangulations

So far we have given experimental evidence and some analytical considerations showing that low Mach number calculations on triangular finite volume cells do not face the *accuracy problem*, whereas this is the very case for grids with rectangular cells. In this section we focus on the question, given a primary grid of triangular cell, what is the behaviour of the first-order Roe scheme on the corresponding dual grid, see [38] for a definition of such grids. For simplicity we restrict our analysis to a dual grid consisting of regular hexagons as depicted in Figure 10.6.

We focus once more on the artificial viscosity terms of order $\mathcal{O}_S(1/M)$ for the transport of horizontal momentum ρu:

$$f_{id}^{\rho u}\left[\mathcal{O}_S(1/M)\right] = -\frac{1}{2}n_x^{id}\rho\hat{a}_{id}\mathbf{n}^{id}\cdot(\mathbf{u}_d - \mathbf{u}_i),$$

$$f_{ie}^{\rho u}\left[\mathcal{O}_S(1/M)\right] = -\frac{1}{2}n_x^{ie}\rho\hat{a}_{ie}\mathbf{n}^{ie}\cdot(\mathbf{u}_e - \mathbf{u}_i),$$

$$f_{ig}^{\rho u}\left[\mathcal{O}_S(1/M)\right] = -\frac{1}{2}n_x^{ig}\rho\hat{a}_{ig}\mathbf{n}^{ig}\cdot(\mathbf{u}_g - \mathbf{u}_i),$$

$$f_{ib}^{\rho u}\left[\mathcal{O}_S(1/M)\right] = -\frac{1}{2}n_x^{ib}\rho\hat{a}_{ib}\mathbf{n}^{ib}\cdot(\mathbf{u}_b - \mathbf{u}_i),$$

where the indices i, b, d, e, g indicate the cells involved. Cells c and f do not contribute to $\mathcal{O}_S(1/M)$ since ρu induces only shear waves in the corresponding Riemann problems. Due to the regularity of the cell we have

$$\mathbf{n}^{ig} = -\mathbf{n}^{id},$$
$$\mathbf{n}^{ib} = -\mathbf{n}^{ie},$$

so that we obtain for the $\mathcal{O}_S(1/M)$ flux difference of the momentum ρu in cell i:

$$\Delta F_i^{\text{visc}}\left[\mathcal{O}_S(1/M)\right] = \frac{1}{2}\rho\left\{n_x^{id}\hat{a}_{id}\mathbf{n}^{id}\cdot(\mathbf{u}_d - 2\mathbf{u}_i + \mathbf{u}_g) + n_x^{ie}\hat{a}_{ie}\mathbf{n}^{ie}\cdot(\mathbf{u}_e - 2\mathbf{u}_i + \mathbf{u}_b)\right\}. \tag{10.14}$$

This equation states that there is a numerical viscosity of order $\mathcal{O}_S(1/M)$ if at least one of two constellations occur in the velocity field: either there is a non-zero curvature (in the sense of finite differences) of the velocity in the direction of the axis along the cells b, i and e – or in the direction of the axis along the cells d, i and g.

It is easy to see that a pair of parallel edges in a finite volume cell leads to a viscosity term of the form given in (10.14) of the order $\mathcal{O}_S(1/M)$ for the velocity component perpendicular to theses interfaces.

Evidence from numerical experiments for the failure of first-order upwind schemes on dual grids can be found in Meister [36, 37] and Viozat et al. [60, 20]. In Figure 10.7 the wrong order relation $\mathcal{O}_S(M)$ for the pressure field is presented with the data taken from [60].

Meister used the AUSMD scheme and Viozat et al. the Roe scheme for low Mach number flow around a NACA0012 aerofoil. Both schemes resolve contact and shear waves and were validated to produce accurate results on primary grids with triangular finite volume cells in Chapter 8.

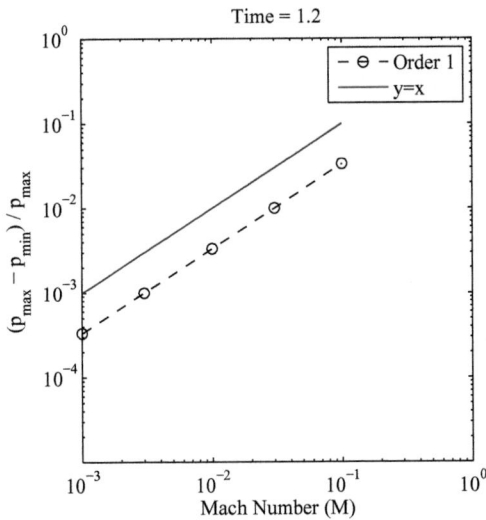

Figure 10.7.: Maximal pressure fluctuation $p_{\text{fluc}} = (p_{\max} - p_{\min})/p_{\max}$ against inflow Mach number for the first-order Roe scheme on a dual grid of a triangulation, cf. [60].

11. Proof for special triangulations

In the last chapters we have seen that the artificial viscosity terms depend on the cell geometry. For cells with parallel edges, the artificial viscosity of order $\mathcal{O}_S(1/M)$ contains second order differences of the flow velocity. This has two possible consequences: either this term vanishes leading to an unphysical velocity field; or the presence of this term creates an unphysical first-order pressure fluctuation.

For triangular finite volume cells, the jumps of the normal velocity at cell interfaces enter the artificial viscosity term, but do not form (standard) second order differences. In numerical experiments these jumps were shown to vanish for $M \to 0$ and fluctuations in the first-order pressure $p^{(1)}$ were never observed. In this chapter we present a rigorous proof that this is indeed the case.

To be precise, we show in this chapter for the first-order Roe scheme in the low Mach number regime:

- $p^{(0)} = \text{const}$ – the pressure of leading order is constant in space,
- $p^{(1)} = \text{const}$ – the pressure of first order is constant in space,
- $\Delta U^{(0)} = 0$ – the leading-order velocity component normal to a cell edge does not jump, and
- $f = \mathcal{O}_S(n)$ – the degrees of freedom f for the velocity components (u, v) under the jump constraint $\Delta U^{(0)} = 0$ is of the same order as the number of grid cells n.

The assumptions used in the analysis are:

- triangular finite volume cells derived from Cartesian grid cells by introducing a diagonal as an additional cell edges,
- a steady flow field, i.e. all temporal derivatives are taken to be zero, and
- $\rho^{(0)} = \text{const}$ – a constant density of leading order.

We think, that these relations imply that the first-order Roe scheme does not suffer from the accuracy problem under the given assumptions. In addition, we derive constraints on far-field and solid wall boundary conditions.

In the analysis presented, we make use of the semi-discrete asymptotic analysis of the Roe scheme. For ease of reading, the derivation of all asymptotic equations was put into Appendix A.5; cross references to these places will be given throughout the text.

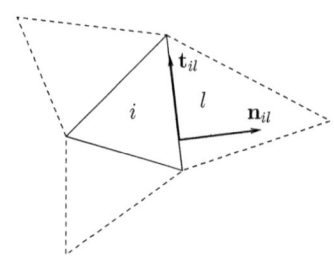

Figure 11.1.: Index notation and local coordinate system for a triangular grid.

Nomenclature We give an overview of the symbols used in this context and refer to Figure 11.1 for illustration. The notation used here is adopted from the asymptotic analysis of the Roe scheme by Viozat in [60].

i or j	index for cell of reference
$\nu(i)$	index set for neighbouring cells
l	index for neighbouring cell
$A_\triangle$	area of the reference cell
il	index for edge between cell i and l
$\mathbf{n}_{il} = (n_x, n_y)_{il}^T$	unit outer normal vector from cell i to l.
$\mathbf{t}_{il} = (-n_y, n_x)_{il}^T$	unit transverse vector from cell i to l.
δ_{il}	length of cell interface il.
$\Delta_{il}\phi = \phi_i - \phi_l$	difference between values of at the reference cell i and its neighbour l.
ϕ_{il}	Roe average of ϕ_i and ϕ_l
$\mathbf{u} = (u, v)^T$	velocity with Cartesian (global) coordinates
$U = \mathbf{u} \cdot \mathbf{n}$	normal component of $\mathbf{u}$
$V = \mathbf{u} \cdot \mathbf{t}$	transverse component of $\mathbf{u}$
Ω	computational domain consisting of finite volume cells

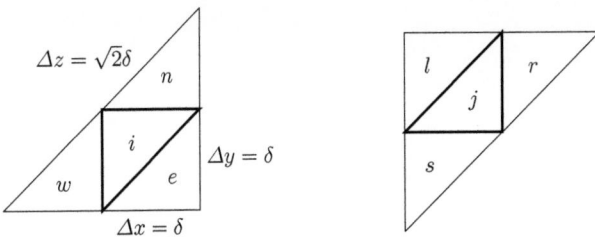

Figure 11.2.: Cell and neighbour cell indices for upper triangle (left) and lower triangle (right) along with edge lengths in the triangulation derived from a Cartesian grid.

11.1. Constancy of the leading-order pressure $p^{(0)}$

The momentum equations corresponding to $\mathcal{O}_S(M^{-1})$ as derived in Appendix A.5 are

$$\sum_{l \in \nu(j)} p_l^{(0)} \cdot (n_x)_{jl} \delta_{jl} = 0 \,, \qquad (M_x^{-1})$$

$$\sum_{l \in \nu(j)} p_l^{(0)} \cdot (n_y)_{jl} \delta_{jl} = 0 \,. \qquad (M_y^{-1})$$

They indicate that the central differences of $p^{(0)}$ in horizontal and in vertical direction vanish separately. To see this, we give some details: the unit outer normal vector $\mathbf{n}$ for an edge and its transverse counterpart $\mathbf{t}$ are

$$\mathbf{n} = \begin{pmatrix} n_x \\ n_y \end{pmatrix}, \qquad \mathbf{t} = \begin{pmatrix} -n_y \\ n_x \end{pmatrix}.$$

If we define the edge vector $\mathbf{d}_{jl}$ to be

$$\mathbf{d}_{jl} = \delta_{jl} \mathbf{t} = \begin{pmatrix} -n_y \delta_{jl} \\ n_x \delta_{jl} \end{pmatrix},$$

where δ_{jl} is the length of the edge separating cell j and l, we can identify in equation (M_x^{-1})

$$(n_x)_{jl} \delta_{jl} = (\mathbf{d}_{jl})_y, \qquad (n_y)_{jl} \delta_{jl} = (\mathbf{d}_{jl})_x \,, \tag{11.1}$$

the vertical and horizontal component of edge vector $\mathbf{d}_{jl}$. For the lower triangle j, compare right part of Figure 11.2, we obtain

$$p_r^{(0)} \Delta y - p_l^{(0)} \Delta y = 0 \qquad \Rightarrow \qquad p_r^{(0)} = p_l^{(0)} \,.$$

In analogy, the equation of vertical momentum gives

$$p_l^{(0)} = p_s^{(0)} \,.$$

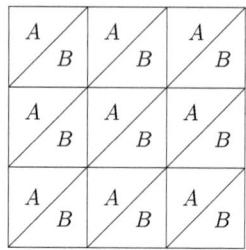

Figure 11.3.: Two-value structure of the leading-order pressure $p^{(0)}$ as a result of the asymptotic momentum equations of order $\mathcal{O}_S(M^{-1})$.

With the corresponding equations for the upper triangle, we see that the central differences of $p^{(0)}$ vanish. This constraint restricts $p^{(0)}$ to form a two-value structure as depicted in Figure 11.3.

The equation for $p^{(0)}$ in its steady form as derived in Appendix A.5 is given by

$$\frac{\gamma-1}{2}\frac{1}{A_\triangle}\sum_{l\in\nu(j)}\frac{h_{jl}^{(0)}}{a_{jl}^{(0)}}\Delta_{jl}p^{(0)}\delta_{jl} = 0 . \tag{P^0}$$

Let us assume, without loss of generality, that $p_j^{(0)} = A > B$. Then all pressure differences in (P^0) satisfy

$$\Delta_{jl}p^{(0)} = A - B > 0 ,$$

so that the sum in (P^0) has to be greater zero, which contradicts the equality to zero. Therefore $A = B$ and the pressure of leading order has to be constant:

$$\boxed{p_i^{(0)} = p_j^{(0)} \quad \text{for} \quad i,j \in \Omega}$$

The constant background pressure $p^{(0)}$ is specified by the boundary conditions, which have to satisfy

$$p_l^{(0)} = p_k^{(0)} , \tag{11.2}$$

for all ghost cells C_l and C_k.

11.2. Constancy of the first-order pressure p$^{(1)}$

To show the constancy of the first-order pressure $p^{(1)}$ it is useful to analyse the equations for the leading order velocity in horizontal and vertical direction, $u^{(0)}$ and $v^{(0)}$, as well as the equation for the first-order pressure $p^{(1)}$. We begin with the general form of these equations and simplify them as much as possible to establish a simple system of linear equations.

11.2.1. Internal cells

At first we analyse the jumps at internal cell interfaces leaving boundary matters to the following section. In the following analysis let us introduce reference quantities, such that

$$p^{(0)} = \frac{1}{\gamma}, \qquad \rho^{(0)} = 1 \quad \Rightarrow \quad [a^{(0)}]^2 = \frac{\gamma p^{(0)}}{\rho^{(0)}} = 1 . \tag{11.3}$$

This is possible since $p^{(0)} = \text{const}$ and, by assumption, $\rho^{(0)} = \text{const}$. To facilitate reading, we omit the superscripts and identify

$$p^{(1)} \rightsquigarrow p, \qquad U^{(0)} \rightsquigarrow U , \tag{11.4}$$

whenever there is no risk of confusing different order terms. In the context of the triangulation derived from a Cartesian grid, two types of grid cells appear, represented with a fat line in Figure 11.2. We call them *upper* and *lower triangle* and allocate different sets of indices to each type. The edge lengths $\Delta x = \Delta y = \delta$ and the length of the diagonal $\sqrt{2}\delta$ drop out of the equations, apart from the $\sqrt{2}$ — the only remnant of the geometric relations in the problem. All time derivatives are omitted because we analyse the steady case.

The equation for $u^{(0)}$ is explicitly derived in Appendix A.5 and reads

$$\underbrace{\frac{d}{dt} u_i^{(0)}}_{\frac{\partial}{\partial t} u^{(0)}} + \underbrace{\frac{1}{\rho_i^{(0)}} \frac{1}{A_\Delta} \sum_{l \in \nu(i)} \frac{p_l^{(1)}(n_x)_{il}}{2} \delta_{il}}_{\frac{1}{\rho^{(0)}} \frac{\partial}{\partial x} p^{(1)}} = -\underbrace{\frac{1}{2} \frac{a_i^{(0)}}{\rho_i^{(0)}} \frac{1}{A_\Delta} \sum_{l \in \nu(i)} \rho_{il}^{(0)} \Delta_{il} U^{(0)} (n_x)_{il} \delta_{il}}_{\frac{1}{2} a^{(0)} \frac{\partial^2}{\partial x^2} u^{(0)} \delta} , \qquad (\text{U}^0)$$

where the underbracing shows the continuous terms of the corresponding modified equation. With the assumptions (11.3) and (11.4), and the geometric relations (11.1) for the cell edges, equation (U^0) can be written for the *upper triangle* as

$$p_e \delta - p_w \delta + \Delta_{ie} U \delta - \Delta_{iw} U \delta = 0 .$$

If we add $-p_i + p_i$ to the LHS, the pressure variables can be transformed into jump variables:

$$\boxed{\Delta_{iw} p - \Delta_{ie} p + \Delta_{ie} U - \Delta_{iw} U = 0 .} \tag{11.5}$$

The analogous equation for the *lower triangle* is

$$\boxed{\Delta_{jl} p - \Delta_{jr} p + \Delta_{jr} U - \Delta_{jl} U = 0 .} \tag{11.6}$$

The equation for $v^{(0)}$ as derived in Appendix A.5 is

$$\underbrace{\frac{d}{dt} v_i^{(0)}}_{\frac{\partial}{\partial t} v^{(0)}} + \underbrace{\frac{1}{\rho_i^{(0)}} \frac{1}{A_\Delta} \sum_{l \in \nu(i)} \frac{p_l^{(1)}(n_y)_{il}}{2} \delta_{il}}_{\frac{1}{\rho^{(0)}} \frac{\partial}{\partial y} p^{(1)}} = -\underbrace{\frac{1}{2} \frac{a_i^{(0)}}{\rho_i^{(0)}} \frac{1}{A_\Delta} \sum_{l \in \nu(i)} \rho_{il}^{(0)} \Delta_{il} U^{(0)} (n_y)_{il} \delta_{il}}_{\frac{1}{2} a^{(0)} \frac{\partial^2}{\partial y^2} v^{(0)} \delta} . \qquad (\text{V}^0)$$

This equation can be transformed in a similar way for the *upper triangle* to

$$\boxed{\Delta_{ie}p - \Delta_{in}p - \Delta_{ie}U + \Delta_{in}U = 0\,,} \tag{11.7}$$

and for the lower triangle to

$$\boxed{\Delta_{js}p - \Delta_{jl}p - \Delta_{js}U + \Delta_{jl}U = 0\,.} \tag{11.8}$$

The equation for $p^{(1)}$, also derived in Appendix A.5, is given by

$$\underbrace{\frac{\mathrm{d}}{\mathrm{d}t}p_i^{(1)}}_{\frac{\partial}{\partial t}p^{(1)}} + \underbrace{\gamma p_i^{(0)} \frac{1}{A_\triangle} \sum_{l \in \nu(i)} \frac{\mathbf{u}_l^{(0)} \cdot \mathbf{n}_{il}}{2}\delta_{il}}_{\gamma p^{(0)} \nabla \cdot \mathbf{u}^{(0)}} = -\underbrace{\frac{1}{2}a_i^{(0)} \frac{1}{A_\triangle} \sum_{l \in \nu(i)} \Delta_{il}p^{(1)}\delta_{il}}_{\frac{1}{2}a^{(0)}\,\nabla^2 p^{(1)}\,\delta}\,. \tag{P^1}$$

We make use of the fact that a closed chain of vectors gives the zero vector, to replace the velocity $\mathbf{u}^{(0)}$ by its corresponding jump variable ΔU:

$$\sum_{l \in \nu(i)} \mathbf{n}_{il}\delta_{il} = 0$$

$$\Rightarrow \sum_{l \in \nu(i)} \mathbf{u}_i^{(0)} \cdot \mathbf{n}_{il}\delta_{il} = 0$$

$$\Rightarrow \sum_{l \in \nu(i)} \mathbf{u}_l^{(0)} \cdot \mathbf{n}_{il}\delta_{il} = \sum_{l \in \nu(i)} \left(\mathbf{u}_l^{(0)} - \mathbf{u}_i^{(0)}\right) \cdot \mathbf{n}_{il}\delta_{il} = -\sum_{l \in \nu(i)} \Delta_{il}U\delta_{il}\,.$$

We furthermore divide equation (P^1) by $a_i^{(0)}$ and make use of

$$\frac{\gamma p^{(0)}}{a^{(0)}} = \rho^{(0)}a^{(0)} = 1\,.$$

The pressure equation can then be written as

$$\sum_{l \in \nu(i)} \left(\Delta_{il}p - \Delta_{il}U\right)\delta_{il} = 0\,.$$

If we apply the special geometry of the *upper triangle* to this equation, we obtain

$$\boxed{\sqrt{2}\Delta_{ie}p + \Delta_{in}p + \Delta_{iw}p - \sqrt{2}\Delta_{ie}U - \Delta_{in}U - \Delta_{iw}U = 0\,.} \tag{11.9}$$

The analogous equation for the *lower triangle* reads

$$\boxed{\sqrt{2}\Delta_{jl}p + \Delta_{js}p + \Delta_{jr}p - \sqrt{2}\Delta_{jl}U - \Delta_{js}U - \Delta_{jr}U = 0\,.} \tag{11.10}$$

The equation for the first-order density $\rho^{(1)}$ is, according to the derivation in Appendix A.5,

$$A_\triangle \frac{\mathrm{d}}{\mathrm{d}t}\rho_i^{(1)} + \frac{1}{2}\sum_{l \in \nu(i)} \left\{\frac{\Delta_{il}p^{(1)}}{a_{il}^{(0)}} + \rho_l^{(0)}\mathbf{u}_l^{(0)} \cdot \mathbf{n}_{il} + |U_{il}^{(0)}|\left(\Delta_{il}\rho^{(0)} - \frac{\Delta_{il}p^{(0)}}{(a_{il}^{(0)})^2}\right)\right\}\delta_{il} = 0\,. \tag{C^1}$$

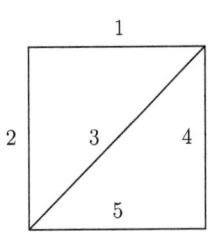 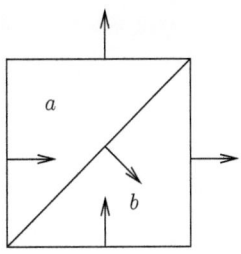

Figure 11.4.: Unified edge indices (left) and orientation of the local coordinate systems (right) for the upper triangle a and lower triangle b.

With the same simplifications used to simplify equation (P^1), along with

$$\rho^{(0)} = \text{const} \quad \Rightarrow \quad \Delta_{il}\rho^{(0)} = 0 \,,$$

we obtain the simple form

$$\sum_{l \in \nu(i)} \left(\Delta_{il} p - \Delta_{il} U \right) \delta_{il} = 0 \,,$$

which is identical to the simplified pressure equation above. The overall system of linear equations consists of three equations per triangle, or six equations per Cartesian cell. To see the relation between these equations, we introduce unified indices for the cell edges as depicted in Figure 11.4, along with unified orientations of the local coordinate systems. With the new indices we redefine the pressure jumps at the cell edges as:

$$\Delta_1 p := p_n - p_i = -\Delta_{in} p \,,$$
$$\Delta_2 p := p_i - p_w = \Delta_{iw} p \,,$$
$$\Delta_3 p := \begin{cases} p_e - p_i = -\Delta_{ie} p \,, \\ p_j - p_l = \Delta_{jl} p \,, \end{cases}$$
$$\Delta_4 p := p_r - p_j = -\Delta_{jr} p \,,$$
$$\Delta_5 p := p_j - p_s = \Delta_{js} p \,.$$

For the jump of the normal velocity ΔU we take into account that this difference was originally defined as outer value minus inner value, i. e. with respect to the outer normal vector. The same holds for the definition of the sign of U. The unified orientations given in Figure 11.4 suggest the following redefinitions:

$$\Delta_1 U := (\mathbf{u}_n - \mathbf{u}_i) \cdot \mathbf{n}_{in} = \Delta_{in} U \,,$$
$$\Delta_2 U := (\mathbf{u}_i - \mathbf{u}_w) \cdot (-\mathbf{n}_{iw}) = (\mathbf{u}_w - \mathbf{u}_i) \cdot \mathbf{n}_{iw} = \Delta_{iw} U \,,$$
$$\Delta_3 U := \begin{cases} (\mathbf{u}_e - \mathbf{u}_i) \cdot \mathbf{n}_{ie} = \Delta_{ie} U \,, \\ (\mathbf{u}_j - \mathbf{u}_l) \cdot (-\mathbf{n}_{jl}) = (\mathbf{u}_l - \mathbf{u}_j) \cdot \mathbf{n}_{jl} = \Delta_{jl} U \,, \end{cases}$$
$$\Delta_4 U := (\mathbf{u}_r - \mathbf{u}_j) \cdot \mathbf{n}_{jr} = \Delta_{jr} U \,,$$
$$\Delta_5 U := (\mathbf{u}_j - \mathbf{u}_s) \cdot (-\mathbf{n}_{js}) = (\mathbf{u}_s - \mathbf{u}_j) \cdot \mathbf{n}_{js} = \Delta_{js} U \,.$$

With the new jump variables $\Delta_1 p$ to $\Delta_5 p$ and $\Delta_1 U$ to $\Delta_5 U$ the system of linear equations (11.7) to (11.10) becomes

$$\Delta_2 p + \Delta_3 p + \Delta_3 U - \Delta_2 U = 0 \,, \tag{Ua}$$
$$-\Delta_3 p + \Delta_1 p - \Delta_3 U + \Delta_1 U = 0 \,, \tag{Va}$$
$$-\sqrt{2}\Delta_3 p - \Delta_1 p + \Delta_2 p - \sqrt{2}\Delta_3 U - \Delta_1 U - \Delta_2 U = 0 \,, \tag{Pa}$$

for the *upper* triangle and

$$\Delta_3 p + \Delta_4 p + \Delta_4 U - \Delta_3 U = 0 \,, \tag{Ub}$$
$$\Delta_5 p - \Delta_3 p - \Delta_5 U + \Delta_3 U = 0 \,, \tag{Vb}$$
$$\sqrt{2}\Delta_3 p + \Delta_5 p - \Delta_4 p - \sqrt{2}\Delta_3 U - \Delta_5 U - \Delta_4 U = 0 \,, \tag{Pb}$$

for the *lower* triangle. The system can be decoupled into equations containing only variables of a single edge, i.e. only two jumps, $(\Delta_i p, \Delta_i U)$ for $i = 1$ to 5, depend on each other. This will be shown next.

In the first step we eliminate $\Delta_3 U$ from all equations using equation (Ua) as pivot equation:

$$\begin{array}{rll}
\text{(Va) + (Ua)} & \Rightarrow & \Delta_1 p + \Delta_2 p + \Delta_1 U - \Delta_2 U = 0 \,, \quad \text{(Va')} \\
\text{(Pa) + }\sqrt{2}\text{ (Ua)} & \Rightarrow & -\Delta_1 p + \alpha\Delta_2 p - \Delta_1 U - \alpha\Delta_2 U = 0 \,, \quad \text{(Pa')} \\
\text{(Ub) + (Ua)} & \Rightarrow & \Delta_2 p + 2\Delta_3 p + \Delta_4 p - \Delta_2 U + \Delta_4 U = 0 \,, \quad \text{(Ub')} \\
\text{(Pb) - (Ua)} & \Rightarrow & -\Delta_2 p - 2\Delta_3 p + \Delta_5 p + \Delta_2 U - \Delta_5 U = 0 \,, \quad \text{(Vb')} \\
\text{(Pb) + }\sqrt{2}\text{ (Ua)} & \Rightarrow & \sqrt{2}\Delta_2 p + 2\sqrt{2}\Delta_3 p - \Delta_4 p + \Delta_5 p & \\
& & \quad -\sqrt{2}\Delta_2 U - \Delta_4 U - \Delta_5 U = 0 \,, \quad \text{(Pb')}
\end{array}$$

where $\alpha = 1 + \sqrt{2}$. In the next step we eliminate $\Delta_3 p$ using (Ub') as pivot equation:

$$\begin{array}{rll}
\text{(Pb') + (Ub')} & \Rightarrow & \Delta_4 p + \Delta_5 p + \Delta_4 U - \Delta_5 U = 0 \,, \quad \text{(Vb'')} \\
\text{(Pb') - }\sqrt{2}\text{ (Ub')} & \Rightarrow & -\alpha\Delta_4 p + \Delta_5 p - \alpha\Delta_4 U - \Delta_5 U = 0 \,. \quad \text{(Pb'')}
\end{array}$$

In the last step we eliminate $\Delta_4 U$ by adding (Pb'') to α times (Vb''):

$$\text{(Pb'') + }\alpha\text{ (Vb'')} \quad \Rightarrow \quad \beta\Delta_5 p - \beta\Delta_5 U = 0 \,, \tag{Pb'''}$$

where $\beta = 1 + \alpha$. The equality $\Delta_5 p = \Delta_5 U$ substituted into (Pb″) gives

$$\Delta_4 p + \Delta_4 U = 0 . \qquad \text{(Vb‴)}$$

If we add (Va′) and (Pa′) we obtain

$$\Delta_2 p - \Delta_2 U = 0 , \qquad \text{(Va‴)}$$

which in turn can be substituted into (Va′) to obtain

$$\Delta_1 p + \Delta_1 U = 0 . \qquad \text{(Va″)}$$

Substituting (Va‴) into (Ua) gives

$$\Delta_3 p + \Delta_3 U = 0 , \qquad \text{(Ua″)}$$

while substituting (Vb‴) into (Ub) gives

$$\Delta_3 p - \Delta_3 U = 0 . \qquad \text{(Ub″)}$$

The last two equations have the unique solution $\Delta_3 p = \Delta_3 U = 0$. We summarise the simplified system of six linear equations:

$$\Delta_1 p + \Delta_1 U = 0 , \qquad (11.11\text{a})$$
$$\Delta_5 p - \Delta_5 U = 0 , \qquad (11.11\text{b})$$
$$\Delta_3 p = 0 , \qquad (11.11\text{c})$$
$$\Delta_3 U = 0 , \qquad (11.11\text{d})$$
$$\Delta_4 p + \Delta_4 U = 0 , \qquad (11.11\text{e})$$
$$\Delta_2 p - \Delta_2 U = 0 . \qquad (11.11\text{f})$$

This system is valid for each pair of triangles inside the entire triangulation, away from boundaries. Note the opposite sign of ΔU in the equation for upper (11.11a) and lower (11.11b) horizontal edges, as well as right (11.11e) and left (11.11f) vertical edges. This allows for some simple statements concerning the jumps in the entire grid.

Diagonal edges

From equation (11.11c) and (11.11d) follows that the pressure of first order $p^{(1)}$ and the normal velocity of leading order $U^{(0)}$ do not jump at any diagonal edge inside the grid.

Internal edges

We consider two neighbouring triangle pairs A and B, as depicted in Figure 11.5, with a common edge i. The corresponding *two systems* of linear equations have only the

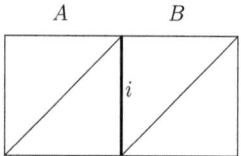

Figure 11.5.: Neighbouring cells with common edge i.

jump variables $\Delta_i p$ and $\Delta_i U$ in common. The equation related to i as the right edge is (11.11e), and for i as the left edge is (11.11f), leading to

$$\Delta_i p + \Delta_i U = 0, \quad \text{(right edge)} \tag{11.12}$$
$$\Delta_i p - \Delta_i U = 0, \quad \text{(left edge)} \tag{11.13}$$

with the unique trivial solution:

$$\Delta_i p = \Delta_i U = 0. \tag{11.14}$$

The same can be shown for vertical edges. The unique solution indicates that the jumps of $p^{(1)}$ and $U^{(0)}$ at all internal edges vanish:

$$\boxed{\Delta_i p = 0, \quad \Delta_i U = 0 \quad \text{for} \quad i \in \Omega/\partial\Omega} \tag{11.15}$$

This property can be given a name:

Definition 11.2.1. *Let* **u** *be a discrete vector field defined in the cells of a triangulation, i.e. every cell is allocated a constant vector.* **u** *is said to be* **zero-jump constrained**, *if the vector components normal to the edges do not jump at these edges.*

As intermediate result we can therefore state a lemma:

Lemma 11.2.2. *Assume a steady solution obtained with the first-order Roe scheme on triangular cells, defined as in Figure 10.1. For* M $\to$ 0 *the solution* **inside the grid** *satisfies:*

- *The pressure of first order $p^{(1)}$ is constant.*
- *The leading-order velocity* $\mathbf{u}^{(0)}$ *is zero-jump constrained.*

So far we have not considered jumps of pressure and velocity at the boundary. This is because the proof relied on a common edge between two pairs of internal finite volume cells, for each of which (11.11a) to (11.11f) are valid. Cells next to the boundary have ghost cell neighbours, for which these equations, derived from the Roe scheme, are not applicable. The values in the ghost cells are derived from boundary conditions, which we now turn to.

11.2.2. Boundary conditions

All boundary conditions we consider are implemented by using ghost cells, which are additional cells that are "glued" to the boundary $\partial\Omega$ of the domain Ω. In this thesis we have investigated external flow with obstacles such as a cylinder or a NACA0012 aerofoil, external flows without obstacles, such as the Gresho vortex, and the inflow of oblique contact layers. We focus here on two sorts of boundary conditions: far-field and solid wall.

Far-field boundaries

For far-field boundaries we "imbed" the flow domain in a steady, uniform flow, which is imposed in the ghost cells. In low Mach number flow it is sensible to assume a constant pressure $p^{(1)}$ of first order in the ghost cells:

$$p^{(1)}_{\text{ghost}} := 1 \,. \tag{11.16}$$

Without global compression or expansion (e. g. for steady flow we analyse here), there is no zero net flux of the leading order velocity $U^{(0)}$ over the boundary:

$$\oint_{\partial\Omega} \mathbf{u}^{(0)} \cdot \mathbf{n} \, dA = 0 \,,$$

or in discrete form:

$$\sum_{i \in \partial\Omega} U_i^{(0)} \delta_i = 0 \,. \tag{11.17}$$

We analyse, whether these conditions avoid jumps of $p^{(1)}$ and $U^{(0)}$ at the boundary. Recall that we identify p with $p^{(1)}$ and U with $U^{(0)}$ to simplify reading in the following.

Constraint on Δp For simplicity we begin with a grid consisting of a pair of triangles, as depicted in Figure 11.6. We impose the pressure values in the ghost cells a to d with the scaling given in (11.3):

$$p_a = p_b = p_c = p_d = 1 \,,$$

while the pressure values in cell α and β, p_α and p_β, are unknown. By definition of the pressure jumps we obtain a simple boundary condition for the $\Delta_i p$'s:

$$\left. \begin{array}{l} \Delta_2 p := p_\alpha - p_b \\ \Delta_3 p := p_\beta - p_\alpha \\ \Delta_5 p := p_\beta - p_d \end{array} \right\} \Rightarrow \Delta_2 p + \Delta_3 p - \Delta_5 p = p_d - p_b = 0 \,.$$

Since the pressure jumps at internal edges are zero by (11.14), the condition is simply

$$\Delta_2 p - \Delta_5 p = 0 \,, \tag{11.18}$$

i. e. we have equality for the two pressure jumps $\Delta_2 p$ and $\Delta_5 p$ at the boundary.

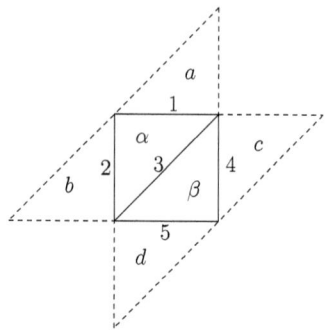

Figure 11.6.: Pair of triangular grid cells α and β with ghost cell neighbours a to d. The numbers 1 to 5 represent edges.

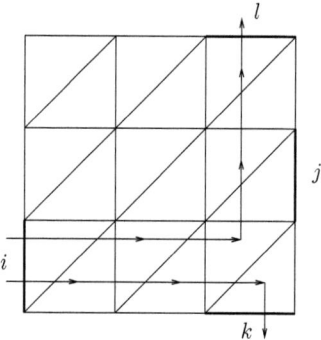

Figure 11.7.: Directed paths connecting boundary edges.

This result can be generalised to a triangulation derived from a Cartesian grid of size $m \times n$ for arbitrary pairs $\Delta_i p$, $\Delta_j p$ of pressure jumps. The sum of pressure jumps at edges on a simple path connecting two boundary edges is always of the form:

$$\pm \Delta_i p \pm \Delta_j p = 0 \,, \tag{11.19}$$

because the pressure jumps "along the way" lie inside the grid and are therefore zero. Due to the definition of positive orientation given in Figure 11.4 (right), the signs have to be specified accordingly. Assume a *directed* path from one ghost cell to another as shown in Figure 11.7. A pressure jump at the boundary edge enters (11.19) with a positive sign if it follows along the local positive direction, otherwise it enters (11.19) with a negative sign. The upper path in Figure 11.7, connecting edge i and l in this sequence, passes both edges in the direction of the local normal vector, therefore:

$$\Delta_i p + \Delta_l p = 0 \,. \tag{11.20}$$

For the lower path, edge k is passed in the opposite direction to the local normal vector and consequently

$$\Delta_i p - \Delta_k p = 0 \,. \tag{11.21}$$

Constraint on ΔU The information on pressure jumps at boundary edges will be useful for showing that all $\Delta_i U$, $i \in \partial \Omega$ are equal. Let us begin with an arbitrary edge i on the left and an arbitrary edge j on the right boundary, as depicted in Figure 11.7. Pressure and velocity jumps at the left edge i and the right edge j are related by (11.12) and (11.13), respectively, while the pressure jumps are related by equation (11.20). We conclude with

$$\left. \begin{array}{rl} \text{left edge:} & \Delta_i p + \Delta_i U = 0 \\ \text{right edge:} & \Delta_j p - \Delta_j U = 0 \\ \text{constraint on } \Delta p: & \Delta_i p + \Delta_j p = 0 \end{array} \right\} \Rightarrow \Delta_i U = \Delta_j U \,, \tag{11.22}$$

that all normal velocity jumps on the left and right boundary are equal. With the same argument all normal velocity jumps on the upper and lower boundary are equal:

$$\left. \begin{array}{rl} \text{upper edge:} & \Delta_l p - \Delta_l U = 0 \\ \text{lower edge:} & \Delta_k p + \Delta_k U = 0 \\ \text{constraint on } \Delta p: & \Delta_l p + \Delta_k p = 0 \end{array} \right\} \Rightarrow \Delta_l U = \Delta_k U \,, \tag{11.23}$$

where l and k are arbitrary edges from the upper and lower boundary, respectively. In the last step we show that the jumps at the left and the upper boundary are equal:

$$\left. \begin{array}{rl} \text{left edge:} & \Delta_i p + \Delta_i U = 0 \\ \text{upper edge:} & \Delta_l p - \Delta_l U = 0 \\ \text{constraint on } \Delta p: & \Delta_i p + \Delta_l p = 0 \end{array} \right\} \Rightarrow \Delta_i U = \Delta_l U \,. \tag{11.24}$$

We call the unique jump of the normal component of the velocity at a boundary edge ΔU and summarise

$$\Delta_i U = \Delta U \quad \text{for} \quad i \in \partial \Omega \,. \tag{11.25}$$

To show that $\Delta U = 0$, we make use of the constraint imposed on the velocity field in the ghost cells – the zero net flux over the boundary.

Zero net flux over the boundary To derive the relation valid for all velocity jumps, we begin, once more, with the simplest triangulation of size 1×1 as depicted in

Figure 11.6. The definitions of the jumps at edge 1 to 5 are:

$\Delta_1 U = v_a - v_\alpha$,

$\Delta_2 U = u_\alpha - u_b$,

$\Delta_3 U = U_\beta - U_\alpha = \dfrac{u_\beta - v_\beta}{\sqrt{2}} - \dfrac{u_\alpha - v_\alpha}{\sqrt{2}} \quad \Rightarrow \quad \sqrt{2}\Delta_3 U = u_\beta - u_\alpha + v_\alpha - v_\beta$,

$\Delta_4 U = u_c - u_\beta$,

$\Delta_5 U = v_\beta - v_d$.

If we sum up all these equations we find

$$\Delta_1 U + \Delta_2 U + \sqrt{2}\Delta_3 U + \Delta_4 U + \Delta_5 U = v_a - v_d + u_c - u_b \ . \tag{11.26}$$

The RHS of (11.26) represents the net flux imposed in the ghost cells, which is assumed to be zero. According to (11.15), all jumps at internal edges vanish: $\Delta_3 U = 0$, and therefore (11.26) is simply

$$\Delta_1 U + \Delta_2 U + \Delta_4 U + \Delta_5 U = 0 \ . \tag{11.27}$$

For the case of an $m \times n$ grid ($2mn$ triangular cells) the sum over all $\Delta_i U$'s reduces to jumps at boundary edges, while the RHS is again the zero net flux of the imposed velocity field in the ghost cells. Therefore we have

$$\sum_{i \in \partial\Omega} \Delta_i U = 0 \ . \tag{11.28}$$

According to (11.25), all jumps at the $2(m+n)$ boundary edges are equal and so (11.28) leads to

$$2(m+n)\Delta U = 0 \quad \Rightarrow \quad \Delta U = 0 \ , \tag{11.29}$$

which states that, under zero net flux boundary conditions for $\mathbf{u}^{(0)}$, there are no jumps of the normal component of the velocity field of leading order at the boundary.

Constancy of $\mathrm{p}^{(1)}$ at the boundary Equations (11.11a) to (11.11f) applied to any boundary edge $i \in \partial\Omega$ can now be simplified to

$$\Delta_i p|_{\partial\Omega} \pm \Delta U|_{\partial\Omega} = 0 \quad \Rightarrow \quad \Delta_i p|_{\partial\Omega} = 0 \ , \tag{11.30}$$

for all pressure jumps at boundary edges. This finishes the proof and we summarise the result in a lemma:

Lemma 11.2.3. *Assume **far-field boundary conditions** as given in* (11.16) *and* (11.17). *Further assume a steady solution obtained with the first-order Roe scheme on triangular cells, defined as in Figure 10.1. For* $\mathrm{M} \to 0$ *the solution at the boundary satisfies:*

- *The pressure of first order $p^{(1)}$ does not jump.*

- *The normal component $U^{(0)}$ of the leading order velocity $\mathbf{u}^{(0)}$ does not jump.*

$$\boxed{\Delta_i p^{(1)} = 0 \ , \quad \Delta_i U^{(0)} = 0 \quad \text{for} \quad i \in \partial\Omega_{\text{far-field}}} \tag{11.31}$$

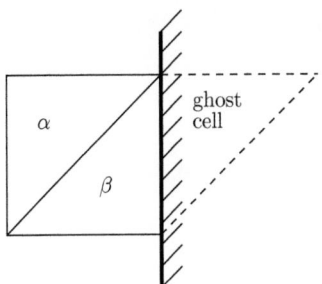

Figure 11.8.: Boundary cell β and ghost cell at a solid wall boundary.

Solid wall boundary conditions

For solid wall boundary conditions in a first-order scheme we impose the following values in the ghost cells and refer to Figure 11.8 for illustration.

$$p_{\text{ghost}} := p_\beta, \qquad \text{(reflexion into ghost cell)}$$
$$\rho_{\text{ghost}} := \rho_\beta, \qquad \text{---''---}$$
$$V_{\text{ghost}} := V_\beta, \qquad \text{---''---}$$
$$U_{\text{ghost}} := -U_\beta, \qquad \text{(reflexion and change of sign)}$$

where U is the component of $\mathbf{u}$ normal to the boundary edge, and V the corresponding transverse component. From the boundary condition imposed on the pressure immediately follows

$$\Delta_i p|_{\partial\Omega} = 0 \,. \tag{11.32}$$

Equations (11.11a) to (11.11f) are now used with (11.32) to find the jumps of the normal velocity at any boundary edge:

$$\Delta_i p|_{\partial\Omega} \pm \Delta_i U|_{\partial\Omega} = 0 \quad \Rightarrow \quad \Delta_i U|_{\partial\Omega} = 0 \,. \tag{11.33}$$

We summarise the result for the solid wall boundary:

Lemma 11.2.4. *Assume **solid-wall boundary conditions** as implemented above. Furthermore assume a steady solution obtained with the first-order Roe scheme on triangular cells, defined as in Figure 10.1. For $M \to 0$ the solution at the boundary satisfies:*

- *The pressure of first order $p^{(1)}$ does not jump.*
- *The normal component $U^{(0)}$ of the leading order velocity $\mathbf{u}^{(0)}$ does not jump.*

$$\boxed{\Delta_i p^{(1)} = 0 \,, \quad \Delta_i U^{(0)} = 0 \quad \text{for} \quad i \in \partial\Omega_{\text{solid wall}}} \tag{11.34}$$

Note on outflow boundaries

If the outflow boundaries are implemented by extrapolating the interior values to the ghost cells, then, obviously, there are no jumps in $U^{(0)}$ or in $p^{(1)}$ across the boundary.

11.3. Degrees of freedom for $\mathbf{u}^{(0)}$

In the last section it was shown that there are, besides the constancy of $p^{(1)}$, no jumps of the normal component of the leading order velocity $\mathbf{u}^{(0)}$ at the cell edges. The question arises, how many degrees of freedom f remain for the velocity field under the zero-jump constraint (11.15). The answer can be found with the help of graph theory. Useful references are [42, 50, 65].

11.3.1. Introduction

In the following analysis we need concepts, which slightly deviate from standard terms used in graph theory. We therefore give the following definitions:

Definition 11.3.1. *The primary grid of finite volume cells induces a* **primary grid graph** *G or simply primary graph in a straightforward manner: grid edges are edges of the graph and where edges meet, we define the vertices of G.*

Definition 11.3.2. *The* **extended dual grid graph** *G^* consists of* **internal vertices** *placed inside each finite volume cell. The ghost cells for boundary conditions give rise to* **external vertices**. *Two vertices are* **neighbours** *if the corresponding primary grid cells have a common edge. External vertices are only connected to their internal neighbour vertices – the corresponding edges are called* **external edges**. **Internal edges** *are the edges connecting internal neighbour vertices with each other.*

Recall that a *dual grid graph*, different to its *extended* counterpart defined above, has only one vertex outside the primary grid graph, which corresponds to the face associated with the outer region. Note that in our context of two-dimensional finite volume grids, the corresponding grid graphs and its extended duals are always *finite, connected and planar*.

11.3.2. Motivating examples

Before elaborating a rigorous derivation for the degrees of freedom, we give a few motivating examples. For a triangulation derived from a Cartesian grid of size $m \times n$ we ask for the degrees of freedom $f_{m,n}$. The following examples suggest a relation to the grid size of the form

$$f_{m,n} = (m-1)(n-1) \ . \tag{11.35}$$

In the following we assume the velocity $\mathbf{u}^{(0)}$ to be given in all ghost cells and ask of the degrees of freedom *inside* the computational domain.

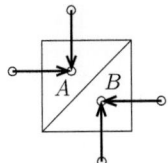

Figure 11.9.: Triangulation derived from a Cartesian grid of size 1 × 1 with arrows indicating the inheritance of normal components from neighbouring ghost cells.

Grid size = 1 × 1 The simplest triangulation derived from a Cartesian grid consists of only two triangles as depicted in Figure 11.9. The thin lines represent the primary grid G of finite volume cells. The thick lines represent the extended dual grid graph G^* and indicate the "inheritance" of velocity components from neighbouring cells. The "inheritance" is founded on the fact, that there are no jumps of the normal components at the cell edges. Thus cell A inherits component u from its left ghost cell neighbour, and component v from the top. Arrow heads underline the direction of inheritance: from cells with fully determined velocity vectors, such as ghost cells, to cells with undetermined components. The values in the ghost cells completely determine the velocity in A, and similarly in B, so that no degree of freedom is left:

$$f_{1,1} = 0,$$

which agrees with (11.35).

Grid size = 1 × 2 It can be verified that, under the zero-jump constraint, all velocity components are determined by the velocity in the ghost cells, so that there is no degree of freedom:

$$f_{1,2} = 0.$$

Grid size = 2 × 2 In Figure 11.10 cells A to F each inherit a component from the neighbouring cell. Note that the top left and lower right cell have completely determined velocities from the boundary and can be neglected in the analysis. The velocity components inherited are called $\tilde{a}$, b, c, $\tilde{d}$, e and f, where we define:

$$a := \sqrt{2}\tilde{a}, \qquad d := \sqrt{2}\tilde{d}.$$

For the edge which is the diagonal of a square, we use the normal directions as indicated in Figure 11.4, so that the corresponding normal component U can be expressed by the Cartesian components u and v:

$$U = \frac{u - v}{\sqrt{2}}.$$

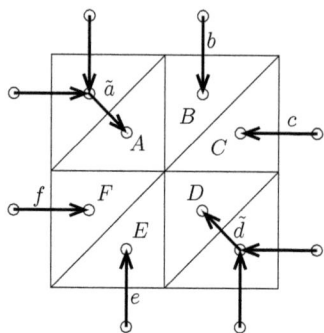

Figure 11.10.: Triangulation derived from a Cartesian grid of size 2 × 2. Arrows indicate the inheritance of normal components of the velocity from neighbouring cells. Cells named A to F and velocities a to f.

Let us choose the y-component of $\mathbf{u}_A$ in cell A as the *free choice* and call it v_A. The normal velocity component relative to the diagonal is $\tilde{a}$ and therefore:

$$\tilde{a} \rightsquigarrow U_A = \frac{u_A - v_A}{\sqrt{2}} \quad \Rightarrow \quad u_A = v_A + a \quad \Rightarrow \quad \mathbf{u}_A = \begin{pmatrix} v_A + a \\ v_A \end{pmatrix},$$

where the arrow "$\rightsquigarrow$" symbolises the "inheritance" of a component from a neighbouring cell. Cell B inherits the y-component from the ghost cell above and the x-component from neighbour A:

$$u_A \rightsquigarrow u_B, \quad b \rightsquigarrow v_B \quad \Rightarrow \quad \mathbf{u}_B = \begin{pmatrix} v_A + a \\ b \end{pmatrix}.$$

Cell C inherits the x-component from the ghost cell to the right: $c \rightsquigarrow u_C$, and the component normal to the diagonal from the neighbouring cell B:

$$U_B \rightsquigarrow U_C \quad \Rightarrow \quad \frac{v_A + a - b}{\sqrt{2}} = \frac{c - v_C}{\sqrt{2}} \quad \Rightarrow \quad \mathbf{u}_C = \begin{pmatrix} c \\ -a + b + c - v_A \end{pmatrix}.$$

Cell D inherits the vertical velocity component from cell C: $v_C = -a+b+c-v_A \rightsquigarrow v_D$ and the component normal to the diagonal from the boundary related value $\tilde{d}$:

$$\frac{d}{\sqrt{2}} \rightsquigarrow U_D = \frac{u_D - v_D}{\sqrt{2}} = \frac{u_D + a - b - c + v_A}{\sqrt{2}}$$

$$\Rightarrow u_D = -a + b + c + d - v_A \quad \Rightarrow \quad \mathbf{u}_D = \begin{pmatrix} -a + b + c + d - v_A \\ -a + b + c - v_A \end{pmatrix}. \quad (11.36)$$

We now approach cell D the other way round via cell F and E. Cell F inherits the vertical component from cell A: $v_A \rightsquigarrow v_F$ and the horizontal component from the ghost cell to the left: $f \rightsquigarrow u_F$. The velocity in F is therefore given by

$$\mathbf{u}_F = \begin{pmatrix} f \\ v_A \end{pmatrix}.$$

Cell E inherits the vertical component from the ghost cell below: $e \rightsquigarrow v_E^*$, and the component normal to the diagonal edge from cell F:

$$U_F \rightsquigarrow U_E \quad \Rightarrow \quad \frac{f - v_A}{\sqrt{2}} = \frac{u_E - e}{\sqrt{2}} \quad \Rightarrow \quad u_E = e + f - v_A$$

$$\Rightarrow \quad \mathbf{u}_E = \begin{pmatrix} e + f - v_A \\ e \end{pmatrix}.$$

Cell D inherits the horizontal velocity component from E: $e + f - v_A \rightsquigarrow u_D$, while the component normal to the diagonal is again inherited across the boundary:

$$\frac{d}{\sqrt{2}} \rightsquigarrow U_D = \frac{u_D - v_D}{\sqrt{2}} = \frac{e + f - v_A - v_D}{\sqrt{2}} \quad \Rightarrow \quad v_D = -d + f - e - v_A$$

$$\Rightarrow \quad \mathbf{u}_D = \begin{pmatrix} e + f - v_A \\ -d + e + f - v_A \end{pmatrix}. \tag{11.37}$$

The two derivations of $\mathbf{u}_D$, given in (11.36) for the path $ABCD$ and in (11.37) for the path $AFED$, should result in the same vector:

$$\mathbf{u}_D^{ABCD} = \mathbf{u}_D^{AFED} \quad \Rightarrow \quad \begin{pmatrix} -a + b + c + d - v_A \\ -a + b + c - v_A \end{pmatrix} = \begin{pmatrix} e + f - v_A \\ -d + e + f - v_A \end{pmatrix}.$$

These are two dependent equations, which are independent of the arbitrary choice v_A, and can be summarised to:

$$\boxed{a + e + f = b + c + d}$$

This constraint simply states that the flow over the boundary of the hexagon consisting of the cells A to F has to be divergence-free. The fact that the result is independent of v_A can be verified for other choices of a velocity component. It shows that, under the zero-jump constraint, only a single component in the hexagon can be chosen arbitrarily without leading to a contradiction:

$$f_{2,2} = 1,$$

which again agrees with Equation (11.35).

Grid size = 3×4 Using the example of a grid of size 3×4, we show the general approach to the problem with concepts from graph theory. In Figure 11.11 a) to h) we illustrate how the construction of the extended dual grid graph G^* can be associated with the degrees of freedom. With the thin lines we represent G. The edges of G^* are the thick lines. Recall they represent the "inheritance" of a velocity component from

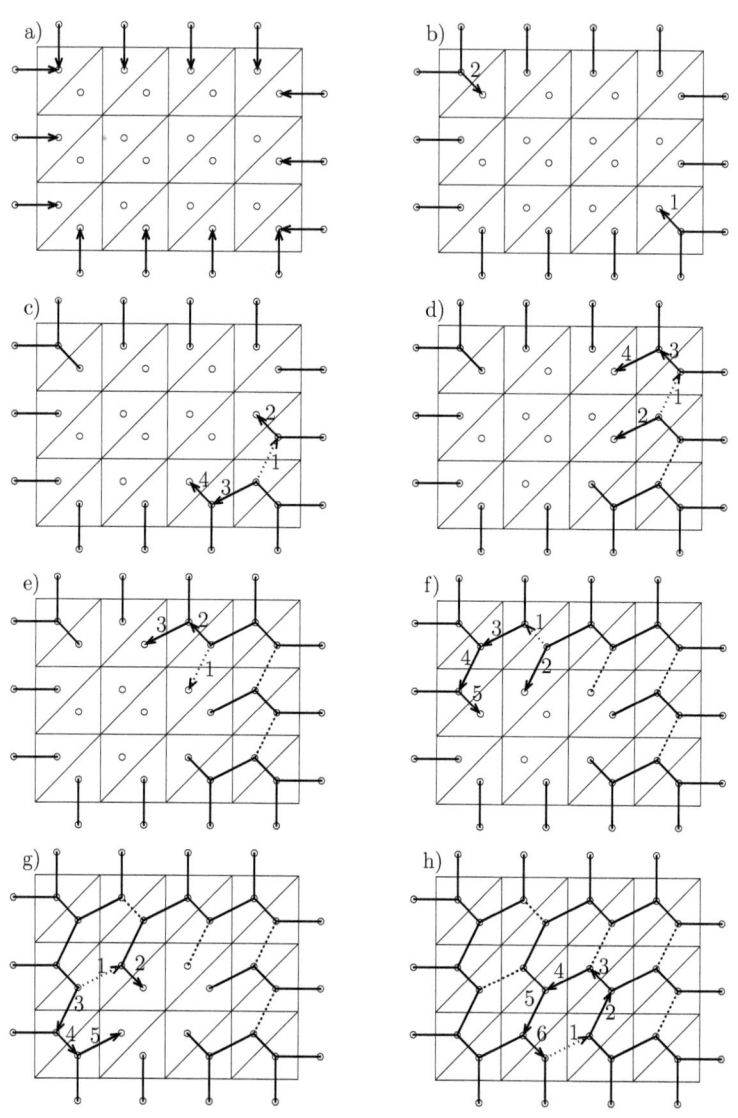

Figure 11.11.: Grid of size 3 × 4 with normal velocities "inherited" from neighbour cells.

a neighbouring cell due to the fact, that there are no jumps of the normal velocity component at a cell edge. We emphasise the direction of inheritance in each step by an arrow head. The sequence of construction is indicated in each subfigure by numbers next to a new edge.

Subfigure a) shows the inheritance from the ghost cells.

If a cell has inherited two velocity components from neighbouring cells, then its velocity vector is fully determined and the cell itself can pass on a normal velocity component to the remaining neighbour. Translated into the graph construction this means: once a vertex in G^* has two edges, we are allowed to draw the missing third edge to the neighbour vertex, as depicted in subfigure b).

If the process of building G^* stalls, there is a free choice to specify an additional velocity component in a cell with only one fixed velocity component. After this "intervention" its velocity vector is fully determined and the inheritance starts again. We represent such a choice of a velocity component inside a cell by a dotted line to one of its neighbours. In subfigure c) we first draw the dotted line 1, and than the other lines 2,3 and 4.

This process can be continued until the entire extended dual graph is complete. Counting the dotted lines will give us the number of velocity components that have been chosen during the construction. It is an upper bound on the number of degrees of freedom. In the 3×4-example shown in Figure 11.11 we obtain six degrees of freedom, which agrees with (11.35):

$$f_{3,4} = (3-1)(4-1) = 6 \ .$$

11.3.3. Graph theoretic analysis

Using the construction principle presented above, we can now approach a general derivation for the degrees of freedom for $\mathbf{u}^{(0)}$ inside the grid. Imagine some dotted lines are given right from the start. Let us call them *fixed edges*. In terms of *degrees of freedom* this means that all free velocity components have been set right from the beginning. If the construction does not stall until G^* is complete, there is no need to specify a further velocity component. In this case we call the graph *constructable*. How many fixed edges are at least necessary for such a graph to be constructable? The answer agrees with the degrees of freedom we look for.

Instead of *constructing* the extended dual grid graph, we delete it, which ends up as the same: let T be a subgraph of G^* with the edges representing the *fixed edges* above missing in T. If the two vector components in a vertex are determined, we are allowed to delete this vertex. The passing on of a component to the remaining neighbour is symbolised by deleting the corresponding edge. Thus, a vertex of degree 1 has inherited already two components and can be deleted together with the adherent edge. These principles can be summarised in the following rule:

Definition 11.3.3 (Rule for deletion). *A vertex has to be deleted if it has degree 1.*

Since we want to know the smallest number of edges missing in T, i.e. we look for some sort of maximal graph. The following definition specifies this idea:

Definition 11.3.4. *We call a subgraph T of G^* **deletable**, if it can be deleted only by applying the rule of deletion. A subgraph T_n is **maximal deletable graph**, if any additional edge makes the resulting subgraph T_{n+1} no longer deletable, where the subscript n indicates the number of edges in the graph.*

The problem of finding the degrees of freedom for a vector field under the zero-jump constraint can now be rephrased: what is the difference between the number of edges in a *maximal deletable* subgraph $T \subset G^*$ and the extended dual grid graph G^*? The answer can be found with the following lemma.

Lemma 11.3.5. *T is a maximal deletable subgraph of an extended dual grid graph G^* if and only if G^* is a spanning tree in G^*.*

Proof. Let us assume, in the first part of the proof, that T is a maximal deletable subgraph of G^*. We have to show, that T is a spanning tree in G^*. Since T is deletable, it does not have cycles. This follows from Definition 11.3.4, because a cycle has only vertices of degree ≥ 2 and there is no way to start deleting such a part of T with the deletion rule. However, if a cycle in G^* cannot be deleted, T itself cannot be (completely) deleted. Furthermore, from T being maximal follows that T has to be connected and contains all vertices of G^*. To see this, let us assume T is not connected. Then T has at least two components, which can be connected by at least *one additional edge* without creating a cycle. The resulting subgraph is still deletable, but then the original graph was not maximal. With the same argument all vertices of G^* must be contained in the subgraph T. Otherwise T would consist of several components.

In the second part, let us assume that T is a spanning tree in G^*. We have to show that T is a maximal deletable subgraph of G^*. Every tree has at least one leaf, cf. [65], i.e. a vertex of degree 1, which can be deleted – and the result is again a tree. We can continue deleting leafs until the entire graph is deleted. Therefore, any spanning tree is *deletable*. To see that any spanning tree T is *maximal* deletable consider: if we add a further edge to T out of the complement $G \setminus T$, we create a cycle and the resulting graph is no longer deletable. Consequently, any spanning tree T in G^* must be maximal deletable subgraph of G^*. □

It is now time to harvest the fruits from the trees. The number of velocity components which can be chosen arbitrarily equals the number of edges missing in the maximal deletable subgraph T compared to G^*. This number is a constant for every grid, since a spanning tree in a graph has always the same number of edges, cf. [65]. We therefore have the following theorem:

Theorem 11.3.6. *Let $\mathbf{u}$ be a discrete vector field defined in the triangular cells of a grid. Let N_e, N_v be the number of edges and vertices in the corresponding extended dual graph G^* and let N_e^{span} be the number of edges of a spanning tree in G^*. The degrees of freedom for $\mathbf{u}$ under the zero-jump constraint is then given by*

$$\boxed{f = N_e - N_e^{\mathrm{span}} = N_e - N_v + 1} \quad (11.38)$$

Proof. We summarise the derivation of this theorem: the degrees of freedom is the minimal number of edges which can be omitted in G^*, so that the resulting subgraph T is still deletable. According to Lemma 11.3.5 such a maximal subgraph must be a spanning tree in G^*. The differences of edges is a fixed number given by $N_e - N_e^{\mathrm{span}}$, which proves the first part of (11.38). The number of edges N_e^{span} in T is related to the number of vertices N_v of G^* by $N_v = N_e^{\mathrm{span}} + 1$, cf. [65]. This proves the second equality in (11.38). □

Let us apply this theorem to a general grid, e.g. an unstructured grid, with n triangles and r boundary edges. Note that we use capital N's for numbers related to the extended dual grid graph G^* and lower case n's for the primary grid graph G. To find the number of edges N_e in G^* we count: n triangles induce n internal vertices of G^*, each of which has degree 3 so that there are $3n$ edges, where the internal edges are counted doubly. We subtract the number of external edges, and obtain $3n - r$ doubly counted internal edges, ergo there are

$$\frac{3}{2}n - \frac{1}{2}r$$

internal edges. Adding r, the number of external edges, gives the total number of edges

$$N_e = \frac{3}{2}n + \frac{1}{2}r . \tag{11.39}$$

To find N_e^{span}, the number of edges in the spanning tree of G^*, we first count the number of vertices in G^*. There are n internal vertices and r external vertices, thus there are

$$N_v = n + r \tag{11.40}$$

vertices in G^*. A spanning tree in G^* has therefore

$$N_e^{\mathrm{span}} = N_v - 1 = n + r - 1$$

edges. If we subtract N_e^{span} from the total number of edges in G^*, we obtain

$$f = \frac{3}{2}n + \frac{1}{2}r - n - r + 1 = \frac{1}{2}n - \frac{1}{2}r + 1 .$$

The result is summarised in the following corollary:

Corollary 11.3.7. *Let a triangulation of a domain have n triangles and r boundary edges. The discrete vector field $\mathbf{u}$ with the zero-jump constraint has*

$$\boxed{f = \frac{1}{2}n - \frac{1}{2}r + 1} \tag{11.41}$$

degrees of freedom.

It is interesting to note, that the degrees of freedom given in (11.41) coincide with the number of internal vertices of the primary grid graph G, called n_v^{in}. To see this, we apply Euler's formula to the extended dual graph G^* and identify faces in G^* with vertices in G.

Theorem 11.3.8 (Euler's formula). *If a finite, connected, planar graph has N_v vertices, N_e edges and N_f faces, then*

$$\boxed{N_v - N_e + N_f = 2}$$

An extended dual grid graph G^* is a finite, connected, planar graph and therefore Eulers's formula applies. The number of vertices N_v of G^* is given by (11.40) and the number of edges by (11.39). Consequently the number of faces is

$$\begin{aligned} N_f &= N_e - N_v + 2 \,, \\ &= \frac{3}{2}n + \frac{1}{2}r - n - r + 2 \,, \\ &= \frac{1}{2}n - \frac{1}{2}r + 2 \,. \end{aligned}$$

To each *internal* vertex of G corresponds a face in G^*. But G^* has one additional face: the region outside the primary grid. Consequently, the number of *internal* vertices n_v^{in} is

$$n_v^{\text{in}} = \frac{1}{2}n - \frac{1}{2}r + 1 \,,$$

which agrees with (11.41). We summarise the finding in the following corollary.

Corollary 11.3.9. *Let a triangulation of a domain have n_e^{in} internal vertices. Then, the degrees of freedom for a discrete vector field under the zero-jump constraint are*

$$\boxed{f = n_e^{\text{in}}}$$

Let us now apply Corollary (11.3.9) to the triangulation derived from the Cartesian grid of size $m \times n$ such as depicted in Figure 11.11. The number of internal vertices is given by

$$n_e^{\text{in}} = (m-1)(n-1) \,,$$

as can be seen by induction. Thus we have the following

Corollary 11.3.10. *Let $\mathbf{u}$ be a discrete vector field defined in the $2mn$ cells of a triangulation derived from a $m \times n$ Cartesian grid with a zero-jump constraint. Then there remain*

$$\boxed{f_{m,n} = (m-1)(n-1)} \tag{11.42}$$

degrees of freedom for the components of $\mathbf{u}$ inside the grid.

Two questions remain to be answered: *First of all, are $f_{m,n}$ enough degrees of freedom? To be more precise, is $f_{m,n}$ of the same order of magnitude as the number of velocity components in high speed flow?* In non-low Mach number flows the number of velocity components that are available for representing a physical velocity field is $4mn$ – two components per triangle – while in the low Mach number flow the degree of freedom is reduced to approximately mn. The degrees of freedom are reduced by a factor of four but are of the same order of magnitude $\mathcal{O}_S(mn)$ and, in this sense, it is enough. Recall that the statement is restricted to the leading order velocity $\mathbf{u}^{(0)}$ in low Mach number flow. Higher-order terms of the velocity are not zero-jump constrained and therefore do not face the problem of having to few degrees of freedom.

Secondly, are the components that can be chosen arbitrarily dispersed throughout the grid or are there, more or less large, regions with all velocity components fixed by the jump constraint? The extended dual graph has elementary cycles positioned around each internal vertex of the primary grid, while the subgraph T has no cycles at all. Therefore, at least one "missing edge", i.e. an edge of $T \backslash G^*$, must be in the vicinity of each internal vertex of the primary grid. This states that the arbitrary choices for the velocity components are well dispersed.

Conclusion

In this chapter we were able to show that the Roe scheme does not suffer from the *accuracy problem* if applied on triangulations derived from Cartesian grids by introducing a diagonal to each square cell. The analysis made use of the steady, discrete, asymptotic equations of the first-order Roe scheme. The density of leading order was assumed constant.

The accuracy problem is characterised by a pressure field of the (unphysical) order $\mathcal{O}_S(M)$, which originates in a numerical viscosity of the wrong order $\mathcal{O}_S(\Delta x/M)$. In the analysis it was shown that on triangular cells the artificial viscosity of this order is determined by the jumps of the normal components of the leading-order velocity at the cell interfaces and that these jumps vanish as $M \to 0$. At the same time the pressure jumps $\Delta p^{(1)}$ were shown to be zero, leading to a constant pressure $p^{(1)}$ of this order. Once the calculation converges to a steady solution, these properties are assumed by $p^{(1)}$ and $\mathbf{u}^{(0)}$, and the *accuracy problem* is circumvented.

In the last part of the proof the zero-jump constraint on the leading-order velocity was shown to leave enough degrees of freedom for the velocity field to represent the physics of the flow.

Summary and Outlook

Concerning Part I

In the first part of this thesis we investigated the numerical behaviour of various first-order upwind schemes in the low Mach number regime in a one-dimensional context. Despite consistency, the numerical behaviour between schemes such as Roe and HLL deviates extremely on a given grid: while the Roe scheme approximates inviscid potential flow, the HLL scheme produces pressure fields typical of *creeping flow* – a flow type known from viscous flow with small Reynolds numbers. We applied the modified equation approach to the upwind schemes, which allowed us to relate the physical convection in the flow to the artificial dissipation introduced in the numerical scheme. The introduction of their ratio – called *numerical Reynolds number* in analogy to its counterpart in viscous flow – allows for a classification of upwind schemes: if the *numerical Reynolds number* is of the order $\mathcal{O}_S(1/\Delta x)$, the artificial damping is "well tuned" relative to the physical transport – independent of the Mach number. The explicit Roe scheme is shown to be a member of this class of *asymptotically consistent* schemes. For the first time in the literature (as far as the author knows) low Mach number flow down to a Mach number of $M = 10^{-14}$ is correctly calculated using the first-order Roe scheme on a grid with a fixed resolution. Note that to date only results were reported of, where the resolution had to be refined for decreasing Mach numbers, cf. Volpe [62].

The class of *asymptotically inconsistent* schemes is defined by a *numerical Reynolds number* of the order $\mathcal{O}_S(M/\Delta x)$ leading to excessive artificial dissipation overweighing the physical convection, which completely changes the type of flow for small Mach numbers comparable to *creeping or Stokes flow*. For the explicit HLL scheme we present numerical results verifying this analogy down to a Mach number of $M = 10^{-12}$. In a dimensional analysis we show the relation between an artificial viscosity of the order $\mathcal{O}_S(M/\Delta x)$ and a pressure of the order $\mathcal{O}_S(M)$. This unphysical pressure field is a typical indicator for the *accuracy problem* and is shown in various pressure-Mach number diagrams in this thesis. This finding is in agreement with the literature, cf. [60, 36]. The one-dimensional analysis can be summarised as follows:

*A first-order upwind scheme can only avoid the **accuracy problem** if all characteristic waves of the local Riemann problem are resolved.*

In practice, first-order upwind schemes are of minor importance. Future effort will therefore be put into investigating the numerical behaviour of second-order upwind schemes, which introduce truncation errors with dispersive effects. At first glance the definition of a *numerical dispersion number*, relating physical convection to artificial dispersion, seems sensible. But first numerical experiments seemed to indicate that the reconstruction worsens the numerical behaviour instead of improving it. First numerical tests with the limiter by Barth & Jespersen [2] for triangular grids produced erroneous pressure oscillations even for moderately small Mach numbers $M_0 \approx 10^{-2}$. In the future we will therefore investigate other limiters that do not have this negative effect, such as the one suggested by Venkatakrishnan [59]. Another approach would be to analyse higher-order schemes on Cartesian grids.

Concerning Part II

Meister shows in [36] and Viozat in [60] with an asymptotic analysis of semi-discrete upwind schemes that pressure fluctuations of the wrong order of magnitude $\mathcal{O}_S(M)$ *can* occur. The derivation is restricted to Cartesian grids. Guillard *et al.* suggest in [19] that the origin of the wrong pressure variations lies in the Riemann problem itself. We point out here, that these results are verified in our thesis for *Cartesian grids*.

For *triangular grids*, however, there is ample numerical evidence shown in this thesis, that the *accuracy problem* does not occur. Thus our results are not in direct contradiction to the literature. We emphasise that also dual grids derived from triangulations lead to the *accuracy problem*. This was shown for a regular grid of hexagons. Numerical evidence of this behaviour can be found by Viozat[60] and Meister [36].

For a special triangulation derived from a Cartesian grid by introducing a diagonal we presented a proof for this phenomenon. Therein we showed that the asymptotic terms of the pressure behave in a similar way as their counterparts in the continuous asymptotic analysis: the pressure of leading and first order, $p^{(0)}$ and $p^{(1)}$, are constant in space. The artificial viscosity of the wrong order $\mathcal{O}_S(\Delta x/M)$ vanishes if there are no jumps of the normal component of the leading-order velocity at the cell interfaces. The absence of these jumps is explicitly derived. In a graph theoretic section it is shown that this jump constraint on the velocity leaves enough degrees of freedom for the velocity field to represent a physical flow. The findings can be summarised as:

A *first-order upwind scheme, which resolves all characteristic waves, does not show the* **accuracy problem** *on triangular finite volume cells.*

In the presented proof we made three major restrictions: steadiness, constant density $\rho^{(0)}$ and a simple cell geometry. In future work answers to the following questions need therefore to be found. What is the behaviour of the Roe scheme in the unsteady case, especially on the time scale of the flow? How stable is the setting with the zero-jump constraint on the leading-order velocity field? In addition, how can the results be generalised to flows with variable-density $\rho^{(0)}$?

The restriction to triangulations derived from Cartesian grids needs to be lifted and the constancy of $p^{(1)}$ to be shown for arbitrary triangular finite volume cells – a result, which was already observed in several numerical experiments on unstructured triangulations.

Appendix

A.1. Roe's approximate Riemann solver

A derivation of Roe's approximate Riemann solver can be found in the textbooks by Laney [32], Leveque [34] or Toro [51]. Here we present an abridged derivation of the major results.

The Roe averages

The linearisation of the flux

$$\mathbf{f}(\mathbf{q}_R) - \mathbf{f}(\mathbf{q}_L) = A(\mathbf{q}_{RL})(\mathbf{q}_R - \mathbf{q}_L)$$

enables us to determine the Roe averages. For the Roe averaged density we set

$$\rho_{RL} = \sqrt{\rho_R \rho_L} ,$$

and obtain for the Roe averaged total enthalpy

$$h_{RL} = \frac{h_R \sqrt{\rho_R} + h_L \sqrt{\rho_L}}{\sqrt{\rho_R} + \sqrt{\rho_L}} .$$

The Roe averaged velocities in x- and y-direction are given by the following term:

$$(u, v)_{RL} = \frac{(u, v)_R \sqrt{\rho_R} + (u, v)_L \sqrt{\rho_L}}{\sqrt{\rho_R} + \sqrt{\rho_L}} ,$$

and the Roe averaged speed of sound is

$$a_{RL}^2 = (\gamma - 1) \left(h_{RL} - \frac{1}{2}(u_{RL}^2 + v_{RL}^2) \right) . \tag{A.43}$$

Constant pressure and density In the analysis it is sometimes advantageous to assume constant pressure p and density ρ. For this reason we derive here, for this assumption, a simplified expression of (A.43) for the Roe averaged speed of sound a_{RL}.

For constant density the Roe averaged specific total enthalpy is simply the arithmetic mean

$$h_{RL} = \frac{h_R + h_L}{2} .$$

With the definition of h as
$$h = \frac{\gamma}{\gamma-1}\frac{p}{\rho} + \frac{1}{2}(u^2+v^2)$$
its Roe average is given by
$$h_{\text{RL}} = \frac{\gamma}{\gamma-1}\frac{p}{\rho} + \frac{1}{2}\left(\frac{u_{\text{R}}^2+u_{\text{L}}^2}{2} + \frac{v_{\text{R}}^2+v_{\text{L}}^2}{2}\right).$$
Using the Roe averaged velocities
$$u_{\text{RL}} = \frac{u_{\text{R}}+u_{\text{L}}}{2} \quad \text{and} \quad v_{\text{RL}} = \frac{v_{\text{R}}+v_{\text{L}}}{2}$$
we obtain for the Roe averaged speed of sound
$$\begin{aligned}a_{\text{RL}}^2 &= (\gamma-1)\left(h_{\text{RL}} - \frac{1}{2}(u_{\text{RL}}^2+v_{\text{RL}}^2)\right)\\ &= (\gamma-1)\left[\frac{\gamma}{\gamma-1}\frac{p}{\rho} + \frac{1}{2}\{\frac{u_{\text{R}}^2+u_{\text{L}}^2}{2} + \frac{v_{\text{R}}^2+v_{\text{L}}^2}{2}\} - \frac{1}{2}\{\frac{(u_{\text{R}}+u_{\text{L}})^2}{4} + \frac{(v_{\text{R}}+v_{\text{L}})^2}{4}\}\right]\\ &= \frac{\gamma p}{\rho} + \frac{1}{2}(\gamma-1)\left\{\frac{u_{\text{R}}^2+u_{\text{L}}^2}{4} - u_{\text{R}}u_{\text{L}} + \frac{v_{\text{R}}^2+v_{\text{L}}^2}{4} - v_{\text{R}}v_{\text{L}}\right\}.\end{aligned}$$
Defining the background velocity as
$$a^2 = \frac{\gamma p}{\rho},$$
we see that the Roe average deviates from this background value by $\mathcal{O}(\text{M})$
$$a_{\text{RL}} = a + \mathcal{O}(\text{M}) \quad \text{as} \quad \text{M} \to 0. \tag{A.44}$$

Eigenvectors and eigenvalues

For the finite volume scheme we need the flux at the origin of the local coordinate system $x=0$, which can be calculated once the approximated state at $x=0$ is known. This Roe averaged state can be obtained by starting from the left or right state and jumping across the waves generated at the local Riemann problem. The flux can then be written in the following forms:
$$\mathbf{f}(\mathbf{q}(0)) \approx A_{\text{RL}}\mathbf{q}(0) = \mathbf{f}(\mathbf{q}_{\text{L}}) + \sum_{i=1}^{4} \mathbf{r}_i \min(0,\lambda_i)\Delta w_i,$$
$$\mathbf{f}(\mathbf{q}(0)) \approx A_{\text{RL}}\mathbf{q}(0) = \mathbf{f}(\mathbf{q}_{\text{R}}) - \sum_{i=1}^{4} \mathbf{r}_i \max(0,\lambda_i)\Delta w_i,$$

or as arithmetic average:

$$\mathbf{f}(\mathbf{q}(0)) \approx A_{\mathrm{RL}}\mathbf{q}(0) = \frac{1}{2}(\mathbf{f}(\mathbf{q}_{\mathrm{R}}) + \mathbf{f}(\mathbf{q}_{\mathrm{L}})) - \sum_{i=1}^{4} \mathbf{r}_i |\lambda_i| \Delta w_i \ .$$

Since the waves are determined in characteristic variables Δw_i we need the right and left eigenvectors of the Jacobian A_{RL}. It can be verified that the following vectors are a linearly independent set of eigenvectors to the corresponding eigenvalues:

$$\mathbf{r}_1 = \begin{pmatrix} 1 \\ u - a \\ v \\ h - au \end{pmatrix} \quad \text{and} \quad \lambda_1 = u - a \ . \tag{A.45a}$$

$$\mathbf{r}_2 = \begin{pmatrix} 1 \\ u \\ v \\ \frac{1}{2}(u^2 + v^2) \end{pmatrix} \quad \text{and} \quad \lambda_2 = u \ , \tag{A.45b}$$

$$\mathbf{r}_3 = \begin{pmatrix} 0 \\ 0 \\ 1 \\ v \end{pmatrix} \quad \text{and} \quad \lambda_3 = u \ , \tag{A.45c}$$

$$\mathbf{r}_4 = \begin{pmatrix} 1 \\ u + a \\ v \\ h + au \end{pmatrix} \quad \text{and} \quad \lambda_4 = u + a \ , \tag{A.45d}$$

In order to obtain the characteristic variables, we need the left eigenvectors given as the rows of the inverse of the transformation matrix $R = [\mathbf{r}_1, \mathbf{r}_2, \mathbf{r}_3, \mathbf{r}_4]$:

$$R^{-1} = L = \begin{pmatrix} \ell_1^T \\ \ell_2^T \\ \ell_3^T \\ \ell_4^T \end{pmatrix} .$$

After some simplifications we find the following system of left eigenvectors:

$$\ell_1^T = \frac{\gamma - 1}{2a^2}(\frac{u^2 + v^2}{2} + \frac{au}{\gamma - 1}, -\frac{a}{\gamma - 1} - u, -v, 1) \ ,$$

$$\ell_2^T = \frac{\gamma - 1}{a^2}(\frac{a^2}{\gamma - 1} - \frac{u^2 + v^2}{2}, u, v, -1) \ ,$$

$$\ell_3^T = (-v, 0, 1, 0) \ ,$$

$$\ell_4^T = \frac{\gamma - 1}{2a^2}(\frac{u^2 + v^2}{2} - \frac{au}{\gamma - 1}, \frac{a}{\gamma - 1} - u, -v, 1) \ .$$

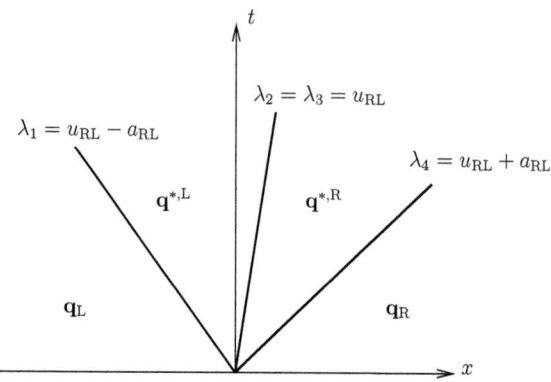

Figure A.12.: Wave structure of the linearised Riemann problem

Characteristic variables

The set of characteristic variables can be obtained from the conservative variables with the transformation
$$\Delta \mathbf{w} = R^{-1} \Delta \mathbf{q} \, ,$$
or for each coordinate
$$\Delta w_i = \ell_i^T \Delta \mathbf{q} \, .$$
The first characteristic variable is given by
$$\Delta w_1 = \frac{1}{2 a_{\text{RL}}} \left(-\rho \Delta u + \frac{\Delta p}{a_{\text{RL}}} \right)$$
and measures the jump across the first wave, called *left-running acoustic wave*. It travels at the speed $u_{\text{RL}} - a_{\text{RL}}$. The second characteristic variable
$$\Delta w_2 = \Delta \rho - \frac{\Delta p}{a_{\text{RL}}^2}$$
describes a jump in the entropy and is therefore referred to as *entropy wave*. The third characteristic variable
$$\Delta w_3 = \rho_{\text{RL}} \Delta v$$
describes the jump of the transverse velocity and is called *shear wave*. The last two waves travel at the speed u_{RL}. The fourth characteristic variable
$$\Delta w_4 = \frac{1}{2 a_{\text{RL}}} \left(\rho_{\text{RL}} \Delta u + \frac{\Delta p}{a_{\text{RL}}} \right)$$
measures the jump travelling at the speed $u_{\text{RL}} + a_{\text{RL}}$ and is called *right-running acoustic wave*.

A.2. Difference operators in 2D

Here we give some differential operators in discrete form for a two-dimensional irregular grid with triangular cells as depicted in Figure A.13. For all derivations we make use of Gauss' theorem in two dimensions for a vector field $\mathbf{w}$ over the domain Ω_i given by a finite volume cell i:

$$\int_{\Omega_i} \nabla \cdot \mathbf{w}\, dA = \oint_{\partial \Omega_i} \mathbf{w} \cdot \mathbf{n}\, ds\ .$$

In each case the line integral can be approximated by a sum over all edges il of grid cell i:

$$\oint_{\partial\Omega_i} \mathbf{w} \cdot \mathbf{n}\, ds \approx \sum_{l \in \nu(i)} \tilde{\mathbf{w}}_{il} \cdot \mathbf{n}_{il} \delta_{il}\ ,$$

so that the mean divergence across cell i can be approximated by

$$\overline{(\nabla \cdot \mathbf{w})}_i \approx \frac{1}{A_\triangle} \sum_{l \in \nu(i)} \tilde{\mathbf{w}}_{il} \cdot \mathbf{n}_{il} \delta_{il}\ ,$$

where $\tilde{\mathbf{w}}_{il}$ is some approximation of $\mathbf{w}$ at the interface il. This approach is now applied to various vector fields to obtain a relation between *discrete differences* emerging from the first-order Roe scheme and continuous *differential operators*.

Discrete divergence

We approximate $\mathbf{u}$ at the cell interface by the average of the neighbouring cells and obtain for the mean divergence over Ω_i corresponding to cell i:

$$\begin{aligned}
\overline{(\nabla \cdot \mathbf{u})}_i &= \frac{1}{A_\triangle} \oint_{\partial \Omega_i} \mathbf{u} \cdot \mathbf{n}\, ds \\
&\approx \frac{1}{A_\triangle} \sum_{l \in \nu(i)} \frac{\mathbf{u}_i + \mathbf{u}_l}{2} \cdot \mathbf{n}_{il} \delta_{il} \\
&= \frac{1}{2} \frac{1}{A_\triangle} \sum_{l \in \nu(i)} \mathbf{u}_i \cdot \mathbf{n}_{il} \delta_{il} + \frac{1}{A_\triangle} \sum_{l \in \nu(i)} \frac{\mathbf{u}_l \cdot \mathbf{n}_{il}}{2} \\
&= \frac{1}{A_\triangle} \sum_{l \in \nu(i)} \frac{\mathbf{u}_l \cdot \mathbf{n}_{il}}{2}\ .
\end{aligned} \qquad (A.46)$$

Note that for any cell holds

$$\sum_{l \in \nu(i)} \mathbf{n}_{il} \delta_{il} = 0\ , \qquad (A.47)$$

since its edges form a closed chain of vectors.

Discrete gradient

The pressure gradient $\nabla p = (p_x, p_y)^T$ has components, which can be written as a divergence:
$$p_x = \nabla \cdot \begin{pmatrix} p \\ 0 \end{pmatrix} \quad \text{and} \quad p_y = \nabla \cdot \begin{pmatrix} 0 \\ p \end{pmatrix}.$$

We approximate the pressure at the cell interface by the arithmetic mean $p_{il} \approx \frac{1}{2}(p_i + p_l)$ and obtain for the mean partial derivatives:

$$\begin{aligned} \overline{(p_x)}_i &= \frac{1}{A_\triangle} \oint \begin{pmatrix} p \\ 0 \end{pmatrix} \cdot \mathbf{n} \, ds \\ &\approx \frac{1}{A_\triangle} \sum_{l \in \nu(i)} \frac{p_i + p_l}{2} (n_x)_{il} \delta_{il} \\ &= \frac{1}{A_\triangle} \sum_{l \in \nu(i)} \frac{p_l (n_x)_{il}}{2} \delta_{il} \end{aligned} \quad (A.48)$$

and similarly

$$\overline{(p_y)}_i \approx \frac{1}{A_\triangle} \sum_{l \in \nu(i)} \frac{p_l (n_y)_{il}}{2} \delta_{il}.$$

Once more we have made use of (A.47) to eliminate p_i.

Discrete Laplacian

The Laplacian of the pressure $\nabla^2 p$ written as a divergence $\nabla \cdot \nabla p$ can be approximated by the following sum

$$\begin{aligned} \overline{(\nabla^2 p)}_i &= \frac{1}{A_\triangle} \oint_{\partial \Omega_i} \nabla p \cdot \mathbf{n} \, ds \\ &\approx \frac{1}{A_\triangle} \sum_{l \in \nu(i)} \frac{\Delta_{li} p}{\bar{\delta}} \delta_{il}, \end{aligned}$$

where we approximated the derivative of the pressure in direction of the outer normal vector of an interface $\mathbf{n}_{il}$ by

$$\nabla p \cdot \mathbf{n}|_{il} \approx \frac{\Delta_{li} p}{\bar{\delta}} \delta_{il},$$

where $\bar{\delta}$ is an averaged edge length. The sum, typically found in the semi-discrete Roe scheme, satisfies therefore

$$\frac{1}{A_\triangle} \sum_{l \in \nu(i)} \Delta_{il} p \delta_{il} \approx \overline{(\nabla^2 p)}_i \bar{\delta}, \quad (A.49)$$

where we used the fact $\Delta_{li} = -\Delta_{il}$.

A.3. Rules for asymptotic expressions

For a detailed introduction to asymptotic analysis we refer to [25]. Here we deduce some basic rules for deriving asymptotic functions for expressions a, b, c which are themselves composed of asymptotic expansions, cf. [36]. An asymptotic 3-term expansion of a quantity f for the asymptotic sequence M^n ($n = 0, 1, 2$) may exist

$$f(M) = f^{(0)} + f^{(1)}M + f^{(2)}M^2 + o(M^2) \quad \text{as} \quad M \to 0.$$

For ease of reading, we suppress a possible dependency of f on space and time in the notation.

Let us assume that f as a function of M can be expanded in a Taylor series about $M = 0$:

$$f(M) = f(0) + \frac{\partial f}{\partial M}(0)M + \frac{1}{2}\frac{\partial^2 f}{\partial M^2}(0)M^2 + \ldots$$

Comparing the asymptotic expansion and the Taylor series returns the corresponding order terms of f:

$$f^{(0)} = f(0),$$

$$f^{(1)} = \frac{\partial f}{\partial M}(0),$$

$$f^{(2)} = \frac{1}{2!}\frac{\partial^2 f}{\partial M^2}(0).$$

In what follows let a, b, c be functions of M (and space and time) which have an asymptotic 3-term expansion with respect to the asymptotic sequence M^n ($n = 0, 1, 2$) and a Taylor expansion with respect to M. We will now investigate the asymptotic expansion of expressions containing some of the functions a, b, c.

Product rule

Let $f = ab$, then we find for $f^{(0)}, f^{(1)}, f^{(2)}$:

$$f^{(0)} = a^{(0)}b^{(0)},$$

$$f^{(1)} = (a'b + ab')(0) = a^{(1)}b^{(0)} + a^{(0)}b^{(1)},$$

$$f^{(2)} = \frac{1}{2}(a''b + 2a'b' + ab'')(0) = a^{(2)}b^{(0)} + a^{(1)}b^{(1)} + a^{(0)}b^{(2)}.$$

Quotient rule

Let $f = a/b$ with $b \neq 0$. Then we find for $f^{(0)}, f^{(1)}, f^{(2)}$:

$$f^{(0)} = \frac{a^{(0)}}{b^{(0)}},$$

$$f^{(1)} = \frac{a^{(1)}b^{(0)} - a^{(0)}b^{(1)}}{(b^{(0)})^2} = \frac{a^{(1)}}{b^{(0)}} - \frac{a^{(0)}b^{(1)}}{(b^{(0)})^2},$$

$$f^{(2)} = \frac{a^{(2)}}{b^{(0)}} - \frac{a^{(0)}b^{(2)}}{(b^{(0)})^2} - \frac{a^{(1)}b^{(1)}}{(b^{(0)})^2} + \frac{a^{(0)}(b^{(1)})^2}{(b^{(0)})^3}.$$

Rule for ab/c

For the expression $f = ab/c$ with $c \neq 0$ we find:

$$f^{(0)} = \frac{a^{(0)}b^{(0)}}{c^{(0)}},$$

$$f^{(1)} = \frac{(a^{(1)}b^{(0)} + a^{(0)}b^{(1)})c^{(0)} - a^{(0)}b^{(0)}c^{(1)}}{(c^{(0)})^2},$$

$$f^{(2)} = \frac{b^{(0)}}{c^{(0)}}a^{(2)} + \frac{1}{c^{(0)}}a^{(1)}b^{(1)} + \frac{a^{(0)}}{c^{(0)}}b^{(2)} - \frac{a^{(0)}b^{(0)}}{(c^{(0)})^2}c^{(2)},$$

$$- \frac{b^{(0)}}{(c^{(0)})^2}a^{(1)}c^{(1)} - \frac{a^{(0)}}{(c^{(0)})^2}b^{(1)}c^{(1)} + \frac{a^{(0)}b^{(0)}}{(c^{(0)})^3}(c^{(1)})^3.$$

A.4. Asymptotic equations for a perfect gas

The asymptotic analysis of the perfect gas law is needed to simplify equations in the asymptotic analysis of the Roe scheme. The total energy per unit volume in dimensional form is given by

$$\hat{\rho}\hat{e} = \frac{\hat{p}}{\gamma - 1} + \frac{1}{2}\hat{\rho}(\hat{u}^2 + \hat{v}^2), \tag{A.50}$$

where the first and second term on the RHS are the internal energy and the kinetic energy per unit volume, respectively. The scaling introduced in Appendix A.5 leads to the nondimensional form of (A.50)

$$\rho e = \frac{p}{\gamma - 1} + \mathrm{M}^2 \frac{1}{2}\rho(u^2 + v^2).$$

With the asymptotic expansions of energy, pressure and velocity the perfect gas law can be split into

$$(\rho e)^{(0)} = \frac{p^{(0)}}{\gamma - 1}, \tag{A.51}$$

$$(\rho e)^{(1)} = \frac{p^{(1)}}{\gamma - 1}, \tag{A.52}$$

$$(\rho e)^{(2)} = \frac{p^{(2)}}{\gamma - 1} + \frac{1}{2}\rho^{(0)}\left([u^{(0)}]^2 + [v^{(0)}]^2\right). \tag{A.53}$$

A.5. Asymptotic equations of the Roe scheme

The analysis presented here is closely related to the one by Viozat et al. in [60]. The major differences are:

- The nomenclature was changed from Cartesian to irregular grids with triangular cells.

- Although it is a semi-discrete analysis, we introduce a time scaling which agrees with an explicit time discretisation: the time steps are $\mathcal{O}_S(M)$ due to the CFL condition. This is reflected in the Strouhal number $Str = 1/M$, appearing as scaling factor for the time derivatives in the conservation laws.

Note that the presented analysis in the thesis is restricted to the steady flow equations, so that this difference will not be of importance.

The 2D Euler equations can be written as:

$$\frac{\partial \mathbf{q}}{\partial t} + \nabla \cdot F(\mathbf{q}) = 0, \tag{A.54}$$

$$F(\mathbf{q}) = \begin{pmatrix} \mathbf{f}(\mathbf{q}) \\ \mathbf{g}(\mathbf{q}) \end{pmatrix}, \tag{A.55}$$

where

$$\mathbf{q} = \begin{bmatrix} \rho \\ \rho u \\ \rho v \\ \rho e \end{bmatrix}, \quad \mathbf{f}(\mathbf{q}) = \begin{bmatrix} \rho u \\ \rho u^2 + p \\ \rho u v \\ (\rho e + p)u \end{bmatrix}, \quad \mathbf{g}(\mathbf{q}) = \begin{bmatrix} \rho v \\ \rho u v \\ \rho v^2 + p \\ (\rho e + p)v \end{bmatrix}. \tag{A.56}$$

Nomenclature Before introducing the Roe scheme, we give an overview of the symbols used in this context and refer to Figure A.13 for illustration.

i	index for cell of reference
$\nu(i)$	index set for neighbouring cells
l	index for neighbouring cell

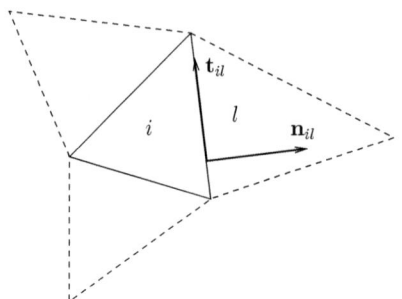

Figure A.13.: index notation and local coordinate system for irregular grid

$A_\triangle$	area of the reference cell
il	index for edge between cell i and l
δ_{il}	length of cell interface il.
$\mathbf{n}_{il} = (n_x, n_y)_{il}^T$	unit outer normal vector from cell i to l.
$\mathbf{t}_{il} = (-n_y, n_x)_{il}^T$	unit transverse vector from cell i to l.
$\Delta_{il}\phi = \phi_i - \phi_l$	difference between values of at the reference cell i and its neighbour l.
ϕ_{il}	Roe average of ϕ_i and ϕ_l
$\mathbf{u} = (u, v)^T$	velocity with Cartesian (global) coordinates
$U = \mathbf{u} \cdot \mathbf{n}$	normal component of $\mathbf{u}$
$V = \mathbf{u} \cdot \mathbf{t}$	transverse component of $\mathbf{u}$

With this notation the characteristic variables can be written in terms of the local coordinate system as:

$$\Delta w_1 = \frac{1}{2a_{il}}\left(\frac{\Delta_{il}p}{a_{il}} - \rho_{il}\Delta_{il}U\right),$$

$$\Delta w_2 = \Delta_{il}\rho - \frac{\Delta_{il}p}{a_{il}^2},$$

$$\Delta w_3 = \rho_{il}\Delta_{il}V,$$

$$\Delta w_4 = \frac{1}{2a_{il}}\left(\frac{\Delta_{il}p}{a_{il}} + \rho_{il}\Delta_{il}U\right).$$

The expressions of the eigenvectors depend on the local coordinate systems of the cell interfaces:

$$\mathbf{r}_1(q_{il}) = \begin{pmatrix} 1 \\ u_{il} - a_{il}(n_x)_{il} \\ v_{il} - a_{il}(n_y)_{il} \\ h_{il} - a_{il}U_{il} \end{pmatrix}, \qquad \mathbf{r}_2(q_{il}) = \begin{pmatrix} 1 \\ u_{il} \\ v_{il} \\ \frac{1}{2}(u_{il}^2 + v_{il}^2) \end{pmatrix},$$

$$\mathbf{r}_3(q_{il}) = \begin{pmatrix} 0 \\ -(n_y)_{il} \\ (n_x)_{il} \\ V_{il} \end{pmatrix}, \qquad \mathbf{r}_4(q_{il}) = \begin{pmatrix} 1 \\ u_{il} + a_{il}(n_x)_{il} \\ v_{il} + a_{il}(n_y)_{il} \\ h_{il} + a_{il}U_{il} \end{pmatrix}.$$

The local eigenvalues are given by

$$\lambda_1 = U_{il} - a_{il}, \quad \lambda_2 = U_{il}, \quad \lambda_3 = U_{il}, \quad \lambda_4 = U_{il} + a_{il}. \tag{A.57}$$

The Roe scheme in two space dimensions can be written as

$$\frac{d}{dt}\mathbf{q}_i + \frac{1}{A_\triangle} \sum_{l \in \nu(i)} \Phi(\mathbf{q}_i, \mathbf{q}_l, \mathbf{n}_{il})\delta_{il} = 0$$

with the Roe flux function

$$\Phi(\mathbf{q}_i, \mathbf{q}_l, \mathbf{n}_{il}) = \frac{F(\mathbf{q}_i) + F(\mathbf{q}_l)}{2} \cdot \mathbf{n}_{il} + \frac{1}{2}\sum_{k=1}^{4} \mathbf{r}_k(\mathbf{q}_{il})|\lambda_k(\mathbf{q}_{il})|\Delta w_k(\Delta_{il}\mathbf{q}).$$

For later use, we give the complete system of equations for the transport of mass density ρ, momentum density ρu and ρv, and the density of total energy ρe:

$$A_\triangle \frac{d}{dt}\rho_i + \frac{1}{2}\sum_{l \in \nu(i)} \rho_l \mathbf{u}_l \cdot \mathbf{n}_{il}\delta_{il}$$
$$+ \frac{1}{2}\sum_{l \in \nu(i)} \left\{ |U_{il}|(\Delta_{il}\rho - \frac{\Delta_{il}p}{a_{il}^2}) + \rho_{il}\frac{U_{il}}{a_{il}}\Delta_{il}U + \frac{\Delta_{il}p}{a_{il}} \right\} \delta_{il} = 0, \tag{A.58}$$

$$A_\triangle \frac{d}{dt}\rho_i u_i + \frac{1}{2}\sum_{l \in \nu(i)} \left\{ \rho_l u_l \mathbf{u}_l \cdot \mathbf{n}_{il} + p_l (n_x)_{il} \right\} \delta_{il}$$
$$+ \frac{1}{2}\sum_{l \in \nu(i)} \left\{ |U_{il}|(\Delta_{il}\rho - \frac{\Delta_{il}p}{a_{il}^2})u_{il} + \rho_{il}\frac{U_{il}}{a_{il}}u_{il}\Delta_{il}U \right\} \delta_{il}$$
$$+ \frac{1}{2}\sum_{l \in \nu(i)} \left\{ -\rho_{il}|U_{il}|(n_y)_{il}\Delta_{il}V + \frac{(Un_x + u)_{il}}{a_{il}}\Delta_{il}p + \rho_{il}a_{il}(n_x)_{il}\Delta_{il}U \right\} \delta_{il} = 0, \tag{A.59}$$

$$A_\triangle \frac{\mathrm{d}}{\mathrm{d}t}\rho_i v_i + \frac{1}{2}\sum_{l\in\nu(i)}\left\{\rho_l v_l \mathbf{u}_l \cdot \mathbf{n}_{il} + p_l(n_y)_{il}\right\}\delta_{il}$$

$$+ \frac{1}{2}\sum_{l\in\nu(i)}\left\{|U_{il}|(\Delta_{il}\rho - \frac{\Delta_{il}p}{a_{il}^2})v_{il} + \rho_{il}\frac{U_{il}}{a_{il}}v_{il}\Delta_{il}U\right\}\delta_{il}$$

$$+ \frac{1}{2}\sum_{l\in\nu(i)}\left\{\rho_{il}|U_{il}|(n_x)_{il}\Delta_{il}V + \frac{(Un_y+v)_{il}}{a_{il}}\Delta_{il}p + \rho_{il}a_{il}(n_y)_{il}\Delta_{il}U\right\}\delta_{il} = 0,$$
(A.60)

$$A_\triangle \frac{\mathrm{d}}{\mathrm{d}t}\rho_i e_i + \frac{1}{2}\sum_{l\in\nu(i)}(\rho_l e_l + p_l)\mathbf{u}_l \cdot \mathbf{n}_{il}\delta_{il}$$

$$+ \frac{1}{2}\sum_{l\in\nu(i)}\left\{|U_{il}|(\Delta_{il}\rho - \frac{\Delta_{il}p}{a_{il}^2})\frac{u_{il}^2+v_{il}^2}{2} + \rho_{il}\frac{U_{il}}{a_{il}}h_{il}\Delta_{il}U\right\}\delta_{il}$$

$$+ \frac{1}{2}\sum_{l\in\nu(i)}\left\{\rho_{il}|U_{il}|V_{il}\Delta_{il}V + \frac{(h+U^2)_{il}}{a_{il}}\Delta_{il}p + \rho_{il}a_{il}U_{il}\Delta_{il}U\right\}\delta_{il} = 0.$$
(A.61)

Nondimensionalisation

We introduce the following scalings in analogy to Subsection 2.2:

l_{ref}	reference length
p_{ref}	reference pressure
ρ_{ref}	reference density
$a_{\mathrm{ref}}^2 = p_{\mathrm{ref}}/\rho_{\mathrm{ref}}$	reference speed of sound
u_{ref}	reference flow velocity
$t_{\mathrm{ref}} = l_{\mathrm{ref}}/a_{\mathrm{ref}}$	reference time
$h_{\mathrm{ref}} = \frac{p_{\mathrm{ref}}}{\rho_{\mathrm{ref}}} = a_{\mathrm{ref}}^2$	reference specific total enthalpy

The dimensionless numbers relevant to this type of flow are Mach and Strouhal number:

$$\mathrm{M} = \frac{u_{\mathrm{ref}}}{a_{\mathrm{ref}}} \quad \text{and} \quad \mathrm{Str} = \frac{l_{\mathrm{ref}}/t_{\mathrm{ref}}}{u_{\mathrm{ref}}} = \frac{1}{\mathrm{M}}.$$

In the following, all equations are sorted by orders of magnitude in M. The nondimensional form of the **continuity equation** is

$$M^0 A_\triangle \frac{d}{dt}\rho_i + \frac{1}{2}\sum_{l\in\nu(i)} \frac{\Delta_{il}p}{a_{il}}\delta_{il}$$

$$+M^1 \frac{1}{2}\sum_{l\in\nu(i)} \left\{\rho_l \mathbf{u}_l \cdot \mathbf{n}_{il} + |U_{il}|(\Delta_{il}\rho - \frac{\Delta_{il}p}{a_{il}^2})\right\}\delta_{il} \qquad (C)$$

$$+M^2 \frac{1}{2}\sum_{l\in\nu(i)} \rho_{il}\frac{U_{il}}{a_{il}}\Delta_{il}U\delta_{il} = 0 \ .$$

A similar procedure leads to the **equation of x-momentum conservation**:

$$M^{-1} \frac{1}{2}\sum_{l\in\nu(i)} p_l(n_x)_{il}\delta_{il}$$

$$+M^0 A_\triangle \frac{d}{dt}\rho_i u_i + \frac{1}{2}\sum_{l\in\nu(i)} \left\{\frac{(Un_x+u)_{il}}{a_{il}}\Delta_{il}p + \rho_{il}a_{il}(n_x)_{il}\Delta_{il}U\right\}\delta_{il}$$

$$+M^1 \frac{1}{2}\sum_{l\in\nu(i)} \left\{\rho_l u_l \mathbf{u}_l \cdot \mathbf{n}_{il} + |U_{il}|(\Delta_{il}\rho - \frac{\Delta_{il}p}{a_{il}^2})u_{il} - \rho_{il}|U_{il}|(n_y)_{il}\Delta_{il}V\right\}\delta_{il}$$

$$+M^2 \frac{1}{2}\sum_{l\in\nu(i)} \rho_{il}\frac{U_{il}}{a_{il}}u_{il}\Delta_{il}U\delta_{il} = 0 \ , \qquad (M_x)$$

and the **equation of y-momentum conservation**:

$$M^{-1} \frac{1}{2}\sum_{l\in\nu(i)} p_l(n_y)_{il}\delta_{il}$$

$$+M^0 A_\triangle \frac{d}{dt}\rho_i v_i + \frac{1}{2}\sum_{l\in\nu(i)} \left\{\frac{(Un_y+v)_{il}}{a_{il}}\Delta_{il}p + \rho_{il}a_{il}(n_y)_{il}\Delta_{il}U\right\}\delta_{il}$$

$$+M^1 \frac{1}{2}\sum_{l\in\nu(i)} \left\{\rho_l v_l \mathbf{u}_l \cdot \mathbf{n}_{il} + |U_{il}|(\Delta_{il}\rho - \frac{\Delta_{il}p}{a_{il}^2})v_{il} - \rho_{il}|U_{il}|(n_x)_{il}\Delta_{il}V\right\}\delta_{il}$$

$$+M^2 \frac{1}{2}\sum_{l\in\nu(i)} \rho_{il}\frac{U_{il}}{a_{il}}v_{il}\Delta_{il}U\delta_{il} = 0 \ . \qquad (M_y)$$

The **equation of energy conservation** in non-dimensional form is

$$\mathrm{M}^{-1} \frac{1}{2} \sum_{l \in \nu(i)} \left\{ |U_{il}|(\Delta_{il}\rho - \frac{\Delta_{il}p}{a_{il}^2}) \frac{u_{il}^2 + v_{il}^2}{2} + \rho_{il}|U_{il}|V_{il}\Delta_{il}V \right\} \delta_{il}$$

$$+\mathrm{M}^0 \, A_\triangle \frac{\mathrm{d}}{\mathrm{d}t}\rho_i e_i + \frac{1}{2} \sum_{l \in \nu(i)} \frac{h_{il}}{a_{il}} \Delta_{il} p \, \delta_{il}$$

$$+\mathrm{M}^1 \frac{1}{2} \sum_{l \in \nu(i)} (\rho_l e_l + p_l) \mathbf{u}_l \cdot \mathbf{n}_{il} \delta_{il} \qquad \qquad \text{(E)}$$

$$+\mathrm{M}^2 \frac{1}{2} \sum_{l \in \nu(i)} \left\{ \frac{U_{il}^2}{a_{il}} \Delta_{il} p + \rho_{il} a_{il} U_{il} \Delta_{il} U + \rho_{il} \frac{U_{il}}{a_{il}} h_{il} \Delta_{il} U \right\} \delta_{il} = 0 \,.$$

Asymptotic equations

As in the continuous asymptotic analysis, we assume for all physical quantities ϕ an asymptotic 3-term expansion

$$\phi = \phi^{(0)} + \mathrm{M} \phi^{(1)} + \mathrm{M}^2 \phi^{(2)} + o(\mathrm{M}^2) \quad \text{as} \quad \mathrm{M} \to 0$$

and insert these expansions in the semi-discrete equations. They read, sorted by powers of the Mach number:

Order M^{-1}:

$$\sum_{l \in \nu(i)} p_l^{(0)} (n_x)_{il} \delta_{il} = 0 \qquad \qquad (\mathrm{M}_x^{-1})$$

$$\sum_{l \in \nu(i)} p_l^{(0)} (n_y)_{il} \delta_{il} = 0 \qquad \qquad (\mathrm{M}_y^{-1})$$

Order M^0:

$$A_\triangle \frac{\mathrm{d}}{\mathrm{d}t} \rho_i^{(0)} + \frac{1}{2} \sum_{l \in \nu(i)} \frac{\Delta_{il} p^{(0)}}{a_{il}^{(0)}} \delta_{il} = 0 \qquad \qquad (\mathrm{C}^0)$$

$$A_\triangle \frac{\mathrm{d}}{\mathrm{d}t} \rho_i^{(0)} u_i^{(0)} + \frac{1}{2} \sum_{l \in \nu(i)} p_l^{(1)} (n_x)_{il} \delta_{il}$$

$$+ \frac{1}{2} \sum_{l \in \nu(i)} \left\{ \frac{(U^{(0)} n_x + u^{(0)})_{il}}{a_{il}^{(0)}} \Delta_{il} p^{(0)} + \rho_{il}^{(0)} a_{il}^{(0)} (n_x)_{il} \Delta_{il} U^{(0)} \right\} \delta_{il} = 0$$

$$(\mathrm{M}_x^0)$$

$$A_\triangle \frac{\mathrm{d}}{\mathrm{d}t}\rho_i^{(0)} v_i^{(0)} + \frac{1}{2} \sum_{l\in\nu(i)} p_l^{(1)} (n_y)_{il}\delta_{il}$$

$$+ \frac{1}{2} \sum_{l\in\nu(i)} \left\{ \frac{(U^{(0)} n_y + v^{(0)})_{il}}{a_{il}^{(0)}} \Delta_{il}p^{(0)} + \rho_{il}^{(0)} a_{il}^{(0)} (n_y)_{il}\Delta_{il} U^{(0)} \right\} \delta_{il} = 0 \quad (\mathrm{M}_y^0)$$

$$A_\triangle \frac{\mathrm{d}}{\mathrm{d}t}\rho_i^{(0)} e_i^{(0)} + \frac{1}{2}\sum_{l\in\nu(i)} \frac{h_{il}^{(0)}}{a_{il}^{(0)}}\Delta_{il}p^{(0)}\delta_{il} = 0 \quad (\mathrm{E}^0)$$

Order M^1:

$$A_\triangle \frac{\mathrm{d}}{\mathrm{d}t}\rho_i^{(1)} + \frac{1}{2}\sum_{l\in\nu(i)}\left\{ \frac{\Delta_{il}p^{(1)}}{a_{il}^{(0)}} + \rho_l^{(0)}\mathbf{u}_l^{(0)}\cdot\mathbf{n}_{il} + |U_{il}^{(0)}|\left(\Delta_{il}\rho^{(0)} - \frac{\Delta_{il}p^{(0)}}{(a_{il}^{(0)})^2}\right)\right\}\delta_{il} = 0 \quad (\mathrm{C}^1)$$

$$A_\triangle \frac{\mathrm{d}}{\mathrm{d}t}(\rho_i u_i)^{(1)} + \frac{1}{2}\sum_{l\in\nu(i)} p_l^{(2)}(n_x)_{il}\delta_{il}$$

$$+ \frac{1}{2}\sum_{l\in\nu(i)} \left\{ \frac{[(U n_x + u)_{il}\Delta_{il}p]^{(1)}}{a_{il}^{(0)}} + [\rho_{il}a_{il}(n_x)_{il}\Delta_{il}U]^{(1)} \right.$$

$$+ \rho_l^{(0)} u_l^{(0)} \mathbf{u}_l^{(0)}\cdot\mathbf{n}_{il} + |U_{il}^{(0)}|\left(\Delta_{il}\rho^{(0)} - \frac{\Delta_{il}p^{(0)}}{(a_{il}^{(0)})^2}\right)u_{il}^{(0)}$$

$$\left. - \rho_{il}^{(0)}|U_{il}^{(0)}|(n_y)_{il}\Delta_{il}V^{(0)} \right\}\delta_{il} = 0 \quad (\mathrm{M}_x^1)$$

$$A_\triangle \frac{\mathrm{d}}{\mathrm{d}t}(\rho_i v_i)^{(1)} + \frac{1}{2}\sum_{l\in\nu(i)} p_l^{(2)}(n_y)_{il}\delta_{il}$$

$$+ \frac{1}{2}\sum_{l\in\nu(i)} \left\{ \frac{[(U n_y + v)_{il}\Delta_{il}p]^{(1)}}{a_{il}^{(0)}} + [\rho_{il}a_{il}(n_y)_{il}\Delta_{il}U]^{(1)} \right.$$

$$+ \rho_l^{(0)} v_l^{(0)} \mathbf{u}_l^{(0)}\cdot\mathbf{n}_{il} + |U_{il}^{(0)}|\left(\Delta_{il}\rho^{(0)} - \frac{\Delta_{il}p^{(0)}}{(a_{il}^{(0)})^2}\right)v_{il}^{(0)}$$

$$\left. + \rho_{il}^{(0)}|U_{il}^{(0)}|(n_x)_{il}\Delta_{il}V^{(0)} \right\}\delta_{il} = 0 \quad (\mathrm{M}_y^1)$$

$$A_\triangle \frac{d}{dt}(\rho_i e_i)^{(1)} + \frac{1}{2}\sum_{l\in\nu(i)}\left\{\frac{(h_{il}\Delta_{il}p)^{(1)}}{a_{il}^{(0)}} + (\rho_l^{(0)}e_l^{(0)} + p_l^{(0)})\mathbf{u}_l^{(0)}\cdot\mathbf{n}_{il}\right\}\delta_{il} = 0 \qquad (\text{E}^1)$$

The terms in brackets with a superscript can be expanded using the rules given in Appendix A.3. The equations corresponding to the order M^2 are not needed for the presented analysis. In the following we extract the equations for pressure and velocity from the equations of mass, momentum and energy.

Evolution equation for $p^{(0)}$

To find the evolution equation for the pressure of leading order we replace in (E^0) the energy density $(\rho e)^{(0)}$ by $p^{(0)}/(\gamma-1)$ using the perfect gas law of leading order (A.51) and obtain

$$\frac{d}{dt}p^{(0)} + \frac{\gamma-1}{2}\frac{1}{A_\triangle}\sum_{l\in\nu(i)}\frac{h_{il}^{(0)}}{a_{il}^{(0)}}\Delta_{il}p^{(0)}\delta_{il} = 0 \ . \qquad (\text{P}^0)$$

Semi-discrete equation for $p^{(1)}$ and $u^{(0)}$

To find the *evolution equation for* $p^{(1)}$ we replace the energy density $(\rho e)^{(1)}$ in Equation (E^1) by the pressure $p^{(1)}$

$$(\rho_i e_i)^{(1)} = \varepsilon_i^{(1)} = \frac{1}{\gamma-1}p_i^{(1)} \ ,$$

cf. Section A.4. We use the fact $p^{(0)} = \text{const}$, proved in Chapter 10, which implies $h^{(0)} = \text{const}$. This simplifies the Roe averages to $h_{il}^{(0)} = h_i^{(0)}$ and $a_{il}^{(0)} = a_i^{(0)}$. Using

$$h_i^{(0)} = \frac{\gamma}{\gamma-1}\frac{p_i^{(0)}}{\rho_i^{(0)}} \quad \text{and} \quad \left(a_i^{(0)}\right)^2 = \frac{\gamma p_i^{(0)}}{\rho_i^{(0)}} \ ,$$

the second term of (E^1) can further be transformed to

$$\frac{(h_{il}\Delta_{il}p)^{(1)}}{a_{il}^{(0)}} = \frac{h_{il}^{(0)}\Delta_{il}p^{(1)} + h_{il}^{(1)}\Delta_{il}p^{(0)}}{a_{il}^{(0)}} = \frac{a_i^{(0)}}{\gamma-1}\Delta_{il}p^{(1)} \ .$$

In the third term we use

$$(\rho_l^{(0)}e_l^{(0)} + p_l^{(0)}) = \frac{\gamma}{\gamma-1}p_l^{(0)}$$

and obtain the

Evolution equation for $p^{(1)}$:

$$\underbrace{\frac{d}{dt}p_i^{(1)}}_{\frac{\partial}{\partial t}p^{(1)}} + \underbrace{\gamma p_i^{(0)} \frac{1}{A_\Delta} \sum_{l \in \nu(i)} \frac{\mathbf{u}_l^{(0)} \cdot \mathbf{n}_{il}}{2} \delta_{il}}_{\gamma p^{(0)} \nabla \cdot \mathbf{u}^{(0)}} = \underbrace{-\frac{1}{2}a_i^{(0)} \frac{1}{A_\Delta} \sum_{l \in \nu(i)} \Delta_{il} p^{(1)} \delta_{il}}_{\frac{1}{2}a^{(0)} \nabla^2 p^{(1)} \delta}. \quad (\text{P}^1)$$

The underbracing shows the terms appearing in the corresponding modified equation. Identifying the differential operators and letting all $\delta_{il} \to 0$, we find the corresponding continuous differential equation for $p^{(1)}$:

$$\frac{\partial}{\partial t}p^{(1)} + \gamma p^{(0)} \nabla \cdot \mathbf{u}^{(0)} = 0, \quad (\text{A.62})$$

which shows the consistency of Equation (P^1) with Equation (2.32) for $p^{(1)}$ we derived in the continuous asymptotic analysis. The second term in (P^1) is the numerical viscosity of the scheme and is $\mathcal{O}(\delta)$ as it should be for a first-order scheme.

We do a similar transformation of (M$_x^0$) to derive the equation for $\mathbf{u}^{(0)}$. Omitting $\Delta_{il} p^{(0)}$ and replacing the Roe averages of constant quantities by their node values $\rho_{il}^{(0)} = \rho_i^{(0)}$ and $a_{il}^{(0)} = a_i^{(0)}$, we obtain the

Evolution equation for $u^{(0)}$:

$$\underbrace{\frac{d}{dt}u_i^{(0)}}_{\frac{\partial}{\partial t}u^{(0)}} + \underbrace{\frac{1}{\rho_i^{(0)}} \frac{1}{A_\Delta} \sum_{l \in \nu(i)} \frac{p_l^{(1)}(n_x)_{il}}{2}\delta_{il}}_{\frac{1}{\rho^{(0)}} \frac{\partial}{\partial x}p^{(1)}} = \underbrace{-\frac{1}{2}\frac{a_i^{(0)}}{\rho_i^{(0)}} \frac{1}{A_\Delta} \sum_{l \in \nu(i)} \rho_{il}^{(0)} \Delta_{il} U^{(0)}(n_x)_{il} \delta_{il}}_{\frac{1}{2}a^{(0)} \frac{\partial^2}{\partial x^2}u^{(0)}\delta}. \quad (\text{U}^0)$$

Beneath the equations, the corresponding continuous terms are given. For completeness we also give the

evolution equation for $v^{(0)}$:

$$\underbrace{\frac{d}{dt}v_i^{(0)}}_{\frac{\partial}{\partial t}v^{(0)}} + \underbrace{\frac{1}{\rho_i^{(0)}} \frac{1}{A_\Delta} \sum_{l \in \nu(i)} \frac{p_l^{(1)}(n_y)_{il}}{2}\delta_{il}}_{\frac{1}{\rho^{(0)}} \frac{\partial}{\partial y}p^{(1)}} = \underbrace{-\frac{1}{2}\frac{a_i^{(0)}}{\rho_i^{(0)}} \frac{1}{A_\Delta} \sum_{l \in \nu(i)} \rho_{il}^{(0)} \Delta_{il} U^{(0)}(n_y)_{il} \delta_{il}}_{\frac{1}{2}a^{(0)} \frac{\partial^2}{\partial y^2}v^{(0)}\delta}. \quad (\text{V}^0)$$

If we let all $\delta_{il} \to 0$, the numerical viscosity of the correct order disappears and we retrieve the x and y component of the evolution equation of $\mathbf{u}^{(0)}$:

$$\frac{\partial}{\partial t}\mathbf{u}^{(0)} + \frac{1}{\rho^{(0)}}\nabla p^{(1)} = 0. \quad (\text{A.63})$$

A.6. Software

The following experimental software was developed to test various first-order upwind schemes on different kinds of grids. All codes use a common library of flux solvers including

- exact Riemann solver (Godunov's original scheme)
- Roe's approximate Riemann solver
- HLL and HLLC
- van Leer splitting
- Steger-Warming splitting
- AUSM, AUSMD

References for these flux solvers are the text book by Toro [51] and the *HYDSOL* software from IAG Stuttgart, which was kindly left to our disposition by Prof. Munz, Dr. Roller and Dr. Dumbser.

CYLFLOW 1

CYLFLOW 1 handles body-fitted, structured grids around a cylinder with quadrilateral finite volume cells. The position of the vertices can be manipulated to allow various grid cell transformations. The grid generation is integrated in the code, which simplifies the study of the transitional behaviour of the schemes from quadrilateral to triangular grid cells. Figure A.14 shows the original regular grid and two kinds of grids obtained therefrom by grid cell transformation. Note that the final grid cells are still quadrilaterals, albeit with well-chosen cell edges reduced in length by a large factor to approximate a triangle. The fat line inside the grid represents a virtual boundary caused by the structured index domain.

CYLFLOW 2

CYLFLOW 2 handles arbitrary, unstructured triangular grids in the *am-fmt-format*. We used the FEM grid generation tool *EMC2* to generate the grid around a NACA0012 aerofoil. The grid generator *CYLGRID* was conceived to generate the grids shown in the middle and on the bottom of Figure A.14 with generic triangular finite volume cells. Note that in *CYLFLOW 1* these grids consist of quadrilaterals with one reduced edge length to approximate a triangle. In Figure A.15 we give examples of an unstructured grid around a cylinder (top) and around a NACA0012 aerofoil (bottom). The coarseness in these example grids is chosen to make the cells better visible. The code is run with finer grid versions.

CYLFLOW 2 was developed to verify the results obtained with the *HYDSOL* code and to test the schemes on structured triangular grids.

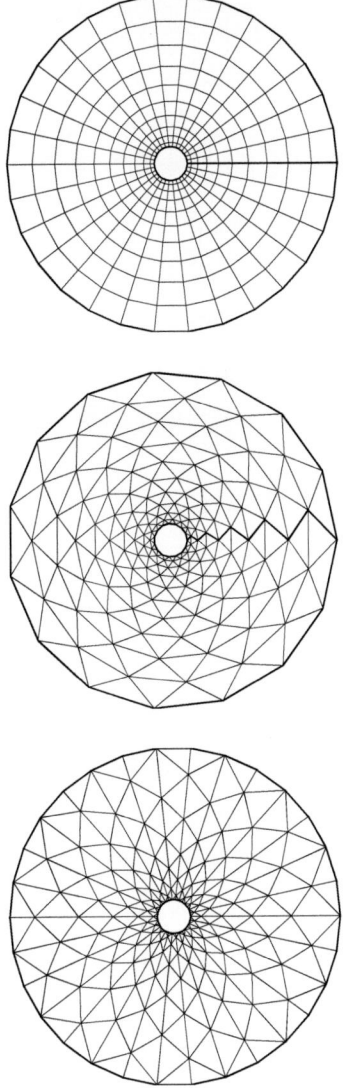

Figure A.14.: Body-fitted structured grids around a cylinder. Top: original polar grid with $n_\phi \times n_r = 30 \times 10$ cells. Middle: quadrilateral cells squeezed in angular direction to approximate triangles. Bottom: quadrilaterals squeezed in radial direction to approximate triangles.

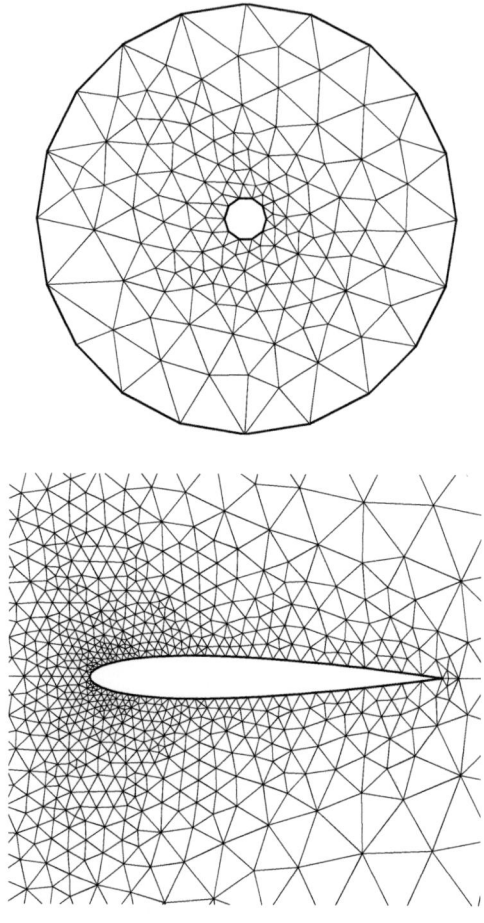

Figure A.15.: Body-fitted unstructured grids around a cylinder with 270 cells (top) and around a NACA0012 aerofoil with 2068 cells (bottom).

CARTFLOW

CARTFLOW was made for calculations on Cartesian grids with and without rectangular obstacles. For comparison a flux formulation and the wave propagation algorithm in the two forms, DCU and CTU, cf. the textbook by Leveque [34], were implemented.

TRIAGFLOW

TRIAGFLOW allows calculations on triangular grids derived from Cartesian grids by introducing the diagonal as additional edge into the square finite volume cell. This allows a direct comparison of the schemes on Cartesian and derived triangular grids.

Notation

List of symbols

a	speed of sound
c	Courant number
c_p	specific heat at constant pressure
c_V	specific heat at constant volume
Δx	cell size in Cartesian grid in x-direction
Δy	cell size in Cartesian grid in y-direction
δ	cell size in a general grid
Δt	size of time step
ε	internal energy per unit volume
e	specific total energy
η	dynamic viscosity
$\mathbf{f}$	flux vector in x-direction
$\mathbf{g}$	flux vector in y-direction
γ	adiabatic index with $\gamma = c_p/c_V$
h	specific total enthalpy
κ	compressibility
λ_i	eigenvalues of the Jacobian A
ℓ_i	left eigenvectors of the Jacobian A
l	length
M	Mach number
$\mathbf{n}_{il}$	normal unit vector on edge between cell i and l
Ω	bounded domain in $\mathbb{R}^d$, where d is the spatial dimension (context dependent)
$\mathcal{O}$	Landau symbol big O
o	Landau symbol small o
$\mathcal{O}_\mathrm{S}$	of the same order of magnitude
Π	stress tensor and

p	pressure
$\mathbf{q}$	vector of conservative variables
$\mathbf{r}_i$	right eigenvector of Jacobian
ρ	density
Re	Reynolds number
Re_{num}	numerical Reynolds number
$\text{Re}_{\text{num}}^{(n)}$	numerical Reynolds Number for the nth characteristic wave
Str	Strouhal number
$(\cdot)_{\text{ref}}$	reference or scaling quantity
t	time
$(\cdot)_t$	first partial derivative with respect to time
$(\cdot)_{tt}$	second partial derivative with respect to time
τ	characteristic damping time
$\mathbf{u}$	flow velocity vector
u, v	x- and y-component of velocity vector $\mathbf{u}$
$\mathbf{w}$	vector of characteristic variables
$(\cdot)_x$	first partial derivative with respect to x
$(\cdot)_{xx}$	second partial derivative with respect to x
$(\cdot)_y$	first partial derivative with respect to y
$(\cdot)_{yy}$	second partial derivative with respect to y
$(\cdot)^{(n)}$	asymptotic term of order n

Conventions

Dimensional quantities are marked with a hat: $\hat{\phi}$.

List of Figures

2.1. Isolines of the pressure fluctuation $\tilde{p} = (p - p_\infty)/p_\infty$ for the creeping flow around a cylinder for a free-stream Mach number $M_\infty = 10^{-3}$. . . . 17

2.2. Isovalues of the pressure fluctuation $\tilde{p}$ for the incompressible potential flow around a cylinder for a free-stream Mach number $M_\infty = 10^{-6}$. . . . 28

3.1. Necessary refinement in terms of cell size δ to simulate flows with decreasing Mach numbers to a given accuracy. Box symbols: first-order Roe scheme. Circle symbols: first-order HLL. 31

4.1. Entropy production due to artificial viscosity and its transport by the explicit first-order Roe scheme for the flow around a cylinder at $M_0 = 10^{-3}$. Shown are the contour lines of the entropy fluctuation $\tilde{s}$. . 47

4.2. Vorticity production and transport for the flow around a cylinder at $M_0 = 10^{-3}$ with the explicit first-order Roe scheme. Shown are the contour lines of the vorticity $\omega = \text{curl } \mathbf{u}$. 49

4.3. Isovalues of pressure for the flow around a cylinder at $M_0 = 10^{-6}$. Left: incompressible potential flow. Right: first-order Roe scheme. . . . 50

4.4. Left: detail of cylinder grid with 9822 cells. Right: pressure coefficient at the cylinder surface of the incompressible potential flow solution (dashed line) and for the explicit first-order Roe scheme (solid line) at $u_\infty/\sqrt{\gamma} = M_0 = 10^{-6}$. 51

4.5. Left: detail of the NACA0012 aerofoil grid with 2068 cells. Right: isolines of pressure for the flow around a NACA0012 aerofoil at $M_0 = 10^{-6}$ with the first-order Roe scheme. 52

4.6. Maximal pressure fluctuation $p_{\text{fluc}} = (p_{\text{max}} - p_{\text{min}})/p_{\text{max}}$ against inflow Mach number for the explicit first-order Roe scheme. 52

5.1. Entropy production due to artificial viscosity by the first-order HLL scheme for the flow around a cylinder at $M_0 = 10^{-3}$. Shown are the contour lines of the entropy fluctuation $\tilde{s}$, cf. Equation (2.6), with a fixed grey scale. 62

5.2. Vorticity production by the first-order HLL scheme for the flow around a cylinder at $M_0 = 10^{-3}$. Shown are the contour lines of the vorticity $\omega = \text{curl } \mathbf{u}$ with a fixed grey scale. 63

5.3. Isovalues of pressure for the flow around a cylinder at $M_0 = 10^{-6}$. Left: incompressible potential flow. Right: explicit first-order HLL. . . . 64

5.4. Maximal pressure fluctuation $p_{\text{fluc}} = (p_{\max} - p_{\min})/p_{\max}$ against the inflow Mach number M for the flow around a cylinder with the first-order HLL scheme. 65

5.5. Isolines of pressure for the flow around a cylinder at $M_0 = 10^{-3}$. Left: creeping flow. Right: first-order HLL. 67

5.6. Surface pressure for the flow around a cylinder at $M_0 = 10^{-3}$. Dashed line: creeping flow. Solid line: first-order HLL. 67

5.7. Steady entropy distribution for the flow around a cylinder at $M_0 = 10^{-3}$. Left: first-order Roe scheme. Right: first-order HLL. 68

5.8. Isolines of pressure for the flow around cylinder at $M_0 = 10^{-6}$. Left: first-order Roe scheme. Right: first-order HLL. 69

5.9. Isolines of pressure for the flow around a NACA0012 aerofoil at $M_0 = 10^{-2}$ with $\alpha = 0°$ (top) and $\alpha = 5°$ (bottom) angle of attack. Left column: first-order Roe scheme. Right column: first-order HLL. . . 70

5.10. Isolines of pressure for the flow around a cylinder at $M_0 = 10^{-6}$. Left: incompressible potential flow. Right: first-order HLLC. 73

6.1. Steady entropy distribution for the flow around a cylinder at $M_0 = 10^{-3}$, obtained with various first-order flux vector splittings. Shown are the contour lines of the entropy fluctuation $\tilde{s}$, cf. Equation (2.6), with different gray scales. 80

6.2. Isolines of pressure for the flow around a cylinder at $M_0 = 10^{-5}$.
Left column: incompressible flow and asymptotically consistent schemes.
Right column: Creeping flow and asymptotically inconsistent schemes. . 82

6.3. Maximal pressure fluctuation $p_{\text{fluc}} = (p_{\max} - p_{\min})/p_{\max}$ against inflow Mach number for the flow around a cylinder obtained with various explicit first-order upwind schemes. 83

7.1. Maximal pressure fluctuation $p_{\text{fluc}} = (p_{\max} - p_{\min})/p_{\max}$ against cell size $\Delta x_n = \Delta x_0/2^n$ for the flow around a cylinder at $M_\infty = 10^{-2}$ for various upwind schemes. 89

7.2. Pressure fluctuation $p_{\text{fluc}} = (p_{\max} - p_{\min})/p_{\max}$ against cell size for the flow around a cylinder at $M_\infty = 10^{-7}$ for various upwind schemes. . . . 90

8.1. Flow around a cylinder with the first-order Roe scheme on a grid with 150x50 cells at $M_0 = 10^{-2}$. Left column: trapezoidal cells converging to triangular cells. Right column: corresponding isolines of pressure converging to potential flow solution. The pressure maximum $p_{max} - 1$ decreases from $4.2 \cdot 10^{-4}$ (top) over $1.6 \cdot 10^{-4}$ (middle) down to $7.7 \cdot 10^{-5}$ (bottom), which is close to the potential flow value of $5.0 \cdot 10^{-5}$. 96

8.2. L_2-error of the pressure as a function of grid cell shape: $\Delta x = \Delta x_0$ represents regular grid cells and $\Delta x / \Delta x_0 \approx 0$ approximately triangular grid cells. Results obtained for the flow around a cylinder with the explicit first-order Roe scheme on a structured grid (cf. Figure 8.3) with $n_\phi \times n_r = 150 \times 50$ cells at $M_0 = 10^{-1}$ down to $M_0 = 10^{-6}$. 97

8.3. Flow around a cylinder with the first-order Roe scheme at $M_0 = 10^{-3}$. Left column: structured, body-fitted grid with 150×50 triangular cells (top). Unstructured, body-fitted grid with 9800 triangular cells (bottom). Right column: isolines of pressure obtained on the corresponding grids. 99

8.4. Left: detail of the structured triangulation around a disc with a layer of quadrilateral cells. Right: isolines of pressure for the flow around a cylinder with the first-order Roe scheme on 150x50 cells at $M_0 = 10^{-3}$. 100

8.5. L_2-error of the pressure as a function of the grid refinement for the flow around a cylinder with the first-order Roe scheme at $M_0 = 10^{-1}$ down to $M_0 = 10^{-6}$. The total number of grid cells in the structured and body-fitted grid are: $n_\phi \times n_r = 30 \times 10$, 60×20, 120×40, 240×80 and 480×160, where n_ϕ is the number of cells around the cylinder and n_r from the cylinder surface to the far-field boundary. 101

8.6. Efficiency study for the flow around a cylinder (150x50 triangular cells) with the explicit first-order Roe scheme on a Pentium 4 with 3.0 GHz. CPU time (left) and number of explicit time steps (right) for convergence to steady state from homogeneous initial conditions. The pressure field settles to steady state after $\mathcal{O}(\log(M))$ time steps. 102

8.7. Triangular grid originating from a Cartesian grid by introducing the diagonal running from the lower left to the upper right corner of the square cell. 103

8.8. Flow around a square with 45° angle of attack at $M_0 = 10^{-3}$ with the explicit first-order Roe scheme. Top: grid with $2 \times 35 \times 35$ triangular cells (left) and isolines of pressure. Middle/Bottom: grid with 50×50 rectangular cells (left) and isolines of pressure obtained with the donor cell upwind method with CFL = 0.45 (middle) and corner upwind transport method at CFL = 0.9 (bottom). 104

8.9. Stagnation point of the flow around a square with $\alpha = 0°$ angle of attack at $M_{\text{inflow}} = 1.0$ with the explicit first-order Roe scheme. Left column: density contours for the grid with triangular cells (top) and square cells (bottom). Right column: pressure contours. In the low Mach number region the density decouples from the pressure field if calculated on rectangular cells. 106

8.10. Mach number contour plots for simulations of a Gresho vortex with the explicit first-order Roe scheme. a) Initial condition. b) Simulation on $2 \times 50 \times 50$ triangular cells at $M_0 = 10^{-1}$ down to $M_0 = 10^{-3}$. c) and d) CTU method on 70×70 Cartesian grid cells at $M_0 = 10^{-1}$ (left) and $M_0 = 10^{-2}$ (right). e) and f) DCU with the same settings. 107

8.11. Inflow of an entropy discontinuity on a grid with $2 \times 50 \times 50$ triangular cells (left column) and 70×70 square cells (right column). The inflow angle varies from $\alpha = 0°$ on top, over $\alpha = 10°$ in the middle to $\alpha = 45°$ in the bottom. The results for different Mach numbers are equal and shown in the same diagram. 110

8.12. Inflow of a shear discontinuity on a grid with $2 \times 50 \times 50$ triangular cells (left column) and 70×70 square cells (right column). The inflow angle is set to $\alpha = 10°$ and the inflow velocity is set according to the given Mach number ranging from $M_0 = 10^{-1}$ down to $M_0 = 10^{-3}$. In the upper half of the grid the inflow Mach number is 1.1 times larger than in the lower half. 112

9.1. Cartesian grid with cell and edge indices. 115
9.2. Grid with parallelogram cells and corresponding indices. 121

10.1. Triangular cells originating from Cartesian grid cells by introducing a diagonal. The global indices (left) are used for the addressing in the numerical algorithm. The local indices (right) are used for the analysis. 125
10.2. Detail of a triangulation based on a Cartesian grid.
Left: edge and cell indices for the analysis of an upper triangle.
Right: local coordinate system for the diagonal edge di. 127
10.3. Right: cell indices for the analysis of the lower triangle.
Left: local coordinate system for the diagonal edge di. 130
10.4. Detail of a Gresho vortex simulation at $M_0 = 10^{-3}$ with the first-order Roe scheme. The grey-scale contour plots show individual components of the piecewise constant velocity field (not interpolated raw data). Left: horizontal components u. Middle: vertical component v. Right: component normal to the diagonal U. 131
10.5. Gresho vortex simulation with the first-order Roe scheme for various Mach numbers. Comparison between the average jump of the velocity component normal to a cell edge, the vorticity ω and the kinetic energy e_{kin} after 1000 time steps. The average jump in u (top), v (middle) and the velocity component normal to the square diagonal U (bottom) are all $\mathcal{O}_S(M^2)$. 132
10.6. Triangulation (dashed line) with corresponding dual grid of regular hexagons (solid line) with corresponding cell indices. 133
10.7. Maximal pressure fluctuation $p_{\text{fluc}} = (p_{\max} - p_{\min})/p_{\max}$ against inflow Mach number for the first-order Roe scheme on a dual grid of a triangulation, cf. [60]. 135

11.1. Index notation and local coordinate system for a triangular grid. 138
11.2. Cell and neighbour cell indices for upper triangle (left) and lower triangle (right) along with edge lengths in the triangulation derived from a Cartesian grid. 139

11.3. Two-value structure of the leading-order pressure $p^{(0)}$ as a result of the asymptotic momentum equations of order $\mathcal{O}_S(M^{-1})$. 140

11.4. Unified edge indices (left) and orientation of the local coordinate systems (right) for the upper triangle a and lower triangle b. 143

11.5. Neighbouring cells with common edge i. 146

11.6. Pair of triangular grid cells α and β with ghost cell neighbours a to d. The numbers 1 to 5 represent edges. 148

11.7. Directed paths connecting boundary edges. 148

11.8. Boundary cell β and ghost cell at a solid wall boundary. 151

11.9. Triangulation derived from a Cartesian grid of size 1×1 with arrows indicating the inheritance of normal components from neighbouring ghost cells. 153

11.10.Triangulation derived from a Cartesian grid of size 2×2. Arrows indicate the inheritance of normal components of the velocity from neighbouring cells. Cells named A to F and velocities a to f. 154

11.11.Grid of size 3×4 with normal velocities "inherited" from neighbour cells. 156

A.12.Wave structure of the linearised Riemann problem 170

A.13.index notation and local coordinate system for irregular grid 176

A.14.Body-fitted structured grids around a cylinder. Top: original polar grid with $n_\phi \times n_r = 30 \times 10$ cells. Middle: quadrilateral cells squeezed in angular direction to approximate triangles. Bottom: quadrilaterals squeezed in radial direction to approximate triangles. 185

A.15.Body-fitted unstructured grids around a cylinder with 270 cells (top) and around a NACA0012 aerofoil with 2068 cells (bottom). 186

Bibliography

[1] A. S. Almgren, J. B. Bell, P. Colella, L. H. Howell, M. L. Welcome, A conservative adaptive projection method for the variable density incompressible Navier-Stokes equations, J. Comput. Phys. 142 (1) (1998) 1–46.

[2] T. Barth, D. Jespersen, The design and application of upwind schemes on unstructured meshes, in: 27th AIAA Aerospace Science Meeting, Reno, NV, 1989, 1989.

[3] P. Batten, N. Clarke, C. Lambert, D. Causon, On the choice of wavespeeds for the HLLC Riemann solver, SIAM J. Sci. Comput. 18 (6) (1997) 1553–1570.

[4] H. Bijl, Computation of flow at all speeds with a staggered scheme, PhD Thesis, Technische Universiteit Delft (1999).

[5] R. B. Bird, W. E. Stewart, E. N. Lightfoot, Transport Phenomena, New York: John Wiley & Sons, 1960.

[6] P. Birken, Numerical simulations of flows at low mach numbers with heat sources, PhD Thesis, Universit"at Kassel (2005).

[7] P. Birken, A. Meister, Stability of preconditioned finite volume schemes at low Mach numbers, BIT 45 (3) (2005) 463–480.

[8] Y.-H. Choi, C. Merkle, The application of preconditioning in viscous flows, J. Comput. Phys. 105 (2) (1993) 207–223.

[9] A. Chorin, A numerical method for solving incompressible viscous flow problems, J. Comput. Phys. 2 (1967) 12–26.

[10] A. Chorin, The numerical solution of the Navier-Stokes equations for an incompressible fluid, Bull. Am. Math. Soc. 73 (1967) 928–931.

[11] I. Demirdzic, Z. Lilek, M. Peric, A collocated finite volume method for predicting flows at all speeds, Int. J. Numer. Methods Fluids 16 (12) (1993) 1029–1050.

[12] P. Deuflhard, F. Bornemann, Numerical mathematics. 2: Ordinary differential equations. (Numerische Mathematik. 2: Gewöhnliche Differentialgleichungen) 2., vollst. überarb. u. erweit. Aufl., de Gruyter Lehrbuch, Berlin: de Gruyter, 2002.

[13] P. Deuflhard, A. Hohmann, Numerische Mathematik. 1: Eine algorithmisch orientierte Einführung. (Numerical mathematics. 1: An algorithmically oriented introduction) 3., überarb. u. erweit. Aufl., de Gruyter Lehrbuch. Berlin: de Gruyter, 2002.

[14] B. Einfeldt, Ein schneller Algorithmus zur Lösung des Riemann-Problems. (An efficient algorithm for the solution to the Riemann problem), Computing 39 (1987) 77–86.

[15] T. Fließbach, Textbook of theoretical physics IV: Statistical physics. (Lehrbuch zur Theoretischen Physik IV: Statistische Physik) 3. Aufl., Spektrum Lehrbuch. Heidelberg: Spektrum Akademischer Verlag., 1999.

[16] K. J. Geratz, Erweiterung eines Godunov-typ-verfahrens für zwei-dimensionale kompressible Strömungen auf die Fälle kleiner und verschwindender Machzahlen, Dissertation, Technische Hochschule Aachen (1997).

[17] P. M. Gresho, On the theory of semi-implicit projection methods for viscous incompressible flow and its implementation via a finite element method that also introduces a nearly consistent mass matrix. I: Theory, Int. J. Numer. Methods Fluids 11 (5) (1990) 587–620.

[18] P. M. Gresho, S. T. Chan, On the theory of semi-implicit projection methods for viscous incompressible flow and its implementation via a finite element method that also introduces a nearly consistent mass matrix. II: Implementation, Int. J. Numer. Methods Fluids 11 (5) (1990) 621–659.

[19] H. Guillard, A. Murrone, On the behavior of upwind schemes in the low Mach number limit. II: Godunov type schemes, Comput. Fluids 33 (4) (2004) 655–675.

[20] H. Guillard, C. Viozat, On the behaviour of upwind schemes in the low Mach number limit, Comput. Fluids 28 (1) (1999) 63–86.

[21] A. Harten, P. D. Lax, B. van Leer, On upstream differencing and Godunov-type schemes for hyperbolic conservation laws, SIAM Rev. 25 (1983) 35–61.

[22] M. H. Holmes, Introduction to perturbation methods. 2nd corrected printing, Texts in Applied Mathematics. 20. New York, NY: Springer., 1998.

[23] K. Karki, S. Patankar, Calculation procedure for viscous incompressible flows in complex geometries, Numer. Heat Transfer 14 (3) (1988) 295–307.

[24] F. Kemm, Carbuncle-freie Flußberechnungen auf der Basis von HLLE und HLLEM, Tech. rep., Brandenburgische Technische Universität Cottbus (2004).

[25] J. K. Kevorkian, J. Cole, Multiple scale and singular perturbation methods, Applied Mathematical Sciences. 114. Berlin: Springer-Verlag., 1996.

[26] S. Klainerman, A. Majda, Compressible and incompressible fluids, Commun. Pure Appl. Math. 35 (1982) 629–651.

[27] R. Klein, Semi-implicit extension of a Godunov-type scheme based on low Mach number asymptotics. I: One-dimensional flow, J. Comput. Phys. 121 (2) (1995) 213–237.

[28] R. Klein, S. Vater, Mathematische Modellierung in der Klimaforschung, Skriptum zur Vorlesung, Wintersemester 2003/04.

[29] N. Kocin, I. Kibel', N. Roze, Theoretische Hydromechanik I, Moskau-Leningrad: OGIZ, Staatsverlag für technisch-theoretische Literatur, 1948.

[30] L. Landau, E. Lifschitz, Textbook of theoretical physics. Vol. V: Statistical physics. Part 1. (Lehrbuch der theoretischen Physik, In deutscher Sprache hrsg. von Paul Ziesche, Band V: Statistische Physik, Teil 1, Übers. aus dem Russ. von Eberhard Jäger, hrsg. von Richard Lenk, E. Lifshits und L. P. Pitaevskij) 8., bericht. Aufl., Frankfurt am Main: H. Deutsch., 1991.

[31] L. Landau, E. Lifschitz, Textbook of theoretical physics, Vol. VI: Hydrodynamics (Lehrbuch der theoretischen Physik, In deutscher Sprache hrsg. von Paul Ziesche, Band VI: Hydrodynamik, In deutscher Sprache hrsg. von Wolfgang Weller, Übers. aus dem Russ. von Adolf Kühnel und Wolfgang Weller) 5., überarb. Aufl., Frankfurt am Main: H. Deutsch, 1991.

[32] C. B. Laney, Computational gasdynamics, Cambridge: Cambridge University Press., 1998.

[33] R. J. Leveque, private communication, ICIAM2007 (Zurich, Switzerland, 2007).

[34] R. J. Leveque, Finite volume methods for hyperbolic problems, Cambridge Texts in Applied Mathematics. Cambridge: Cambridge University Press, 2002.

[35] J. Lighthill, Waves in fluids, Cambridge Mathematical Library. Cambridge: Cambridge University Press, 2001.

[36] A. Meister, Analyse und Anwendung Asymptotik-basierter numerischer Verfahren zur Simulation reibungsbehafteter Strömungen in allen Mach-Zahlbereichen, Habilitationsschrift, Universität Hamburg (2001).

[37] A. Meister, Asymptotic-based preconditioning technique for low Mach number flows, ZAMM, Z. Angew. Math. Mech. 83 (1) (2003) 3–25.

[38] A. e. Meister, J. e. Struckmeier, Hyperbolic partial differential equations. Theory, numerics and applications (with contributions by Michael Junk, Claus-Dieter Munz, Milovan Perić, Demirdžić, Samir Muzaferija, Giovanni Russo and Thomas Sonar), Braunschweig: Vieweg, 2002.

[39] C.-D. Munz, S. Roller, R. Klein, K. Geratz, The extension of incompressible flow solvers to the weakly compressible regime, Comput. Fluids 32 (2) (2003) 173–196.

[40] K. Nerinckx, J. Vierendeels, E. Dick, A pressure-correction algorithm with mach-uniform efficiency and accuracy, Int. J. Numer. Methods Fluids 47 (10-11) (2005) 1205–1211.

[41] K. Nerinckx, J. Vierendeels, E. Dick, Mach-uniformity through the coupled pressure and temperature correction algorithm, J. Comput. Phys. 206 (2) (2005) 597–623.

[42] M. Nitzsche, Graphs for beginners – Around the house of Santa Claus, (Graphen für Einsteiger, Rund um das Haus vom Nikolaus) 2nd corrected ed., Wiesbaden: Vieweg, 2005.

[43] C. W. Oseen, Über die *Stokes*sche Formel und über eine verwandte Aufgabe in der Hydrodynamik (1910), 1911.

[44] J.-H. Park, Ein konservatives MPV-Verfahren zur Simulation der Strömungen in allen Machzahlbereichen, Dissertation, IAG Universität Stuttgart (2003).

[45] P. Roe, Approximate Riemann solvers, parameter vectors, and difference schemes, Journal of Computational Physics 43 (1981) 357–372.

[46] S. Roller, Ein numerisches Verfahren zur Simulation schwach kompressibler Strömungen, Dissertation, IAG Universität Stuttgart (2004).

[47] T. Schneider, N. Botta, K. Geratz, R. Klein, Extension of finite volume compressible flow solvers to multi-dimensional, variable density zero Mach number flows, J. Comput. Phys. 155 (2) (1999) 248–286.

[48] J. Sesterhenn, B. Müller, H. Thomann, On the cancellation problem in calculating compressible low Mach number flows, J. Comput. Phys. 151 (2) (1999) 597–615.

[49] J. H. Spurk, Introduction to the theory of fluid mechanics (Strömungslehre, Einführung in die Theorie der Strömungen) 5. erweiterte Auflage., Berlin: Springer, 2004.

[50] P. Tittmann, Graphentheorie. Eine anwendungsorientierte Einführung, Hanser Fachbuchverlag, 2003.

[51] E. F. Toro, Riemann solvers and numerical methods for fluid dynamics, A practical introduction, 2nd ed., Berlin: Springer, 1999.

[52] L. N. Trefethen, Group velocity in finite difference schemes, SIAM Rev. 24 (1982) 113–136.

[53] L. N. Trefethen, Finite difference and spectral methods for ordinary and partial differential equations, unpublished text, available at http://web.comlab.ox.ac.uk/oucl/work/nick.trefethen/pdetext.html (1996).

[54] E. Turkel, Preconditioned methods for solving the incompressible and low speed compressible equations, J. Comput. Phys. 72 (1987) 277–298.

[55] E. Turkel, Review of preconditioning methods for fluid dynamics, Appl. Numer. Math. 12 (1-3) (1993) 257–284.

[56] E. Turkel, A. Fiterman, B. van Leer, Frontiers of Computational Fluid Dynamics 1994, chap. Preconditioning and the limit of the compressible to the incompressible flow equations for finite difference schemes, CMAS: Computational Methods in Applied Sciences. Chichester: Wiley., 1995.

[57] D. van der Heul, C. Vuik, P. Wesseling, A conservative pressure-correction method for flow at all speeds, Comput. Fluids 32 (8) (2003) 1113–1132.

[58] B. van Leer, W.-T. Lee, P. Roe, Characteristic time-stepping or local preconditioning of the euler equations, AIAA paper 91-1552, 1991.

[59] V. Venkatakrishnan, Convergence to steady state solutions of the Euler equations on unstructured grids with limiters, J. Comput. Phys. 118 (1) (1995) 120–130.

[60] C. Viozat, Implicit upwind schemes for low mach number compressible flows, Tech. rep., Institut National de Recherche en Informatique et en Automatique (INRIA) (1997).

[61] H. Vogel, C. Gerthsen, Physik. Mit 90 Tab. und 105 durchgerechneten Beisp. und 1065 Aufg. mit vollst. Lösungen. (Physics. With 90 tab. and 105 examples and 1065 problems with complete solutions).18., völlig neu bearb. Aufl., Berlin: Springer-Verlag., 1995.

[62] G. Volpe, Performance of compressible flow codes at low Mach numbers, AIAA J. 31 (1) (1993) 49–56.

[63] R. Warming, B. Hyett, The modified equation approach to the stability and accuracy analysis of finite-difference methods, J. Comput. Phys. 14 (1974) 159–179.

[64] P. Wesseling, D. van der Heul, A unified method for compressible and incompressible flows with general equation of state, Toro, E. F. (ed.), Godunov methods. Theory and applications. International conference, Oxford, GB, October 1999. New York, NY: Kluwer Academic/ Plenum Publishers, 1057-1064 (2001) (2001).

[65] D. B. West, Introduction to graph theory. 2nd ed., New Delhi: Prentice-Hall of India. 608 p. , 2005.

[66] J. Zierep, Similarity laws and modeling, M. Dekker, 1971.

[67] J. Zierep, Ähnlichkeitsgesetze und Modellregeln der Strömungslehre, Wissenschaft und Technik, Taschenausgabe, Karlsruhe: G. Braun Verlag , 1972.

[68] D. Zwillinger, Handbook of differential equations, With CD ROM, 3rd ed., San Diego, CA: Academic Press, 1998.

Die VDM Verlagsservicegesellschaft sucht für wissenschaftliche Verlage abgeschlossene und herausragende

Dissertationen, Habilitationen, Diplomarbeiten, Master Theses, Magisterarbeiten usw.

für die kostenlose Publikation als Fachbuch.

Sie verfügen über eine Arbeit, die hohen inhaltlichen und formalen Ansprüchen genügt, und haben Interesse an einer honorarvergüteten Publikation?

Dann senden Sie bitte erste Informationen über sich und Ihre Arbeit per Email an *info@vdm-vsg.de*.

Sie erhalten kurzfristig unser Feedback!

VDM Verlagsservicegesellschaft mbH
Dudweiler Landstr. 99
D - 66123 Saarbrücken

Telefon +49 681 3720 174
Fax +49 681 3720 1749

www.vdm-vsg.de

Die VDM Verlagsservicegesellschaft mbH vertritt

Printed by Books on Demand GmbH, Norderstedt / Germany